AF361925

Environmental Planning and Modeling

Environmental Planning and Modeling

Editor

Christian N. Madu

MDPI • Basel • Beijing • Wuhan • Barcelona • Belgrade • Manchester • Tokyo • Cluj • Tianjin

Editor
Christian N. Madu
Pace University
USA

Editorial Office
MDPI
St. Alban-Anlage 66
4052 Basel, Switzerland

This is a reprint of articles from the Special Issue published online in the open access journal *Sustainability* (ISSN 2071-1050) (available at: https://www.mdpi.com/journal/sustainability/special_issues/environmental_planning).

For citation purposes, cite each article independently as indicated on the article page online and as indicated below:

LastName, A.A.; LastName, B.B.; LastName, C.C. Article Title. *Journal Name* **Year**, *Volume Number*, Page Range.

ISBN 978-3-0365-5101-2 (Hbk)
ISBN 978-3-0365-5102-9 (PDF)

Cover image courtesy of Christian N. Madu.

Contents

About the Editor

Christian N. Madu

Christian N. Madu, PhD is Professor of Management Science at the Lubin School of Business, Pace University, New York, and Professor of Environmental Management and Control at the University of Nigeria. His research interests are environmental planning and applications of management techniques to solve sustainability issues. Prof. Dr. Christian N. Madu is on the editorial board of several journals and has also served as editor-in-chief of major international journals. He has published widely in multidisciplinary fields and is the author/editor of more than 17 books by leading academic publishers. His current interests are chemical safety and security management.

Preface to "Environmental Planning and Modeling"

The focus of this Special Issue reprint book is environmental planning and modeling. Planning is critical in the effective management of the environment and its natural resources. The management of natural resources requires the evaluation of alternative options and selecting those that are optimal. A systemic view of the environment is taken in this Special Issue to ensure that models presented to manage the environment consider the multi-faceted nature of our natural environment as a system.

The rise in the consumption and exploitation of the Earth's finite and limited resources requires that we adopt measures that are sustainable by changing our practices and seeking alternative substitutes. An increase in consumption is often fueled by increasing demands and growing population rates. To meet these demands, production has significantly increased over the years. This increase in production often requires the exploitation of energy and water resources, and land. Consequences abound as we continue to claim more land to build on, thus affecting natural biodiversity, which ultimately impacts the food chain. Deforestation is a problem in many countries as we continue to use forest reserves without measures to replace them. The energy demands to produce the items we consume encourage the use of fossil fuels. These activities lead to the generation of carbon and other greenhouse gases that affect the Earth's atmosphere. Thus, climate change is now more difficult to mitigate as we seek ways to adapt to it. There are also risks associated with our unsustainable activities, such as health hazards, the depletion of resources for future generations, and our inability to cope with climate change.

It is imperative that we take actions that address how we can live in a sustainable world. Such actions require that we develop effective plans to evaluate every available option, apply life cycle assessments in a systemic framework, and make decisions and policies that are supportive of a sustainable environment. Environmental planning enables us to adapt to the needs of our dynamic and changing world and avoid the risks, high costs, and irreparable losses that are associated with poor environmental practices. Modeling of proven effective practices is necessary to standardize the planning framework while still offering an adaptable framework that can be used in situation-specific areas.

The articles in this special edition have already been published online, but are worthy of greater circulation and exposure. The country-specific applications in this book can offer insight into how the models could be adapted in other areas. These articles are summarized in the editorial. The articles present potential areas for future research. They are also useful for application and practice orientation. It is our hope that other researchers can expand on the views espoused in this book. The book will, therefore, be of interest to researchers, students of sustainability and environmental management, policy planners, and decision makers.

Christian N. Madu
Editor

 *sustainability*

Editorial

Editorial on Environmental Planning and Modeling

Christian N. Madu [1,2]

[1] Department of Management & Management Science, Lubin School of Business, Pace University, 1 Pace Plaza, New York, NY 10038, USA; cn_madu@yahoo.com or cmadu@pace.edu
[2] Center for Environmental Management & Control, University of Nigeria, Enugu Campus, Nsukka 410001, Nigeria

Citation: Madu, C.N. Editorial on Environmental Planning and Modeling. *Sustainability* 2022, 14, 9728. https://doi.org/10.3390/su14159728

Received: 1 August 2022
Accepted: 4 August 2022
Published: 8 August 2022

Publisher's Note: MDPI stays neutral with regard to jurisdictional claims in published maps and institutional affiliations.

The aim of this Special Issue is to facilitate environmental decision making by considering the natural environment, as well as social, political, economic, and governance issues. This holistic approach of the natural environment around achieving sustainability offers a win–win for both society and the environment. It is therefore imperative that models developed as decision-supports to enhance environmental policy and decision making consider the vital influences of socio-economic, political, and governance issues in effective decision/policy making. This is even more important as concerns about climate change, food security, and resources limitations are linked with increasing population growth in developing countries. Our focus here was to solicit research papers that are systemic in scope and yet integrative of these factors; building a framework to consider these factors, developing novel methods or models, or applying existing models may be carried out to solve any of the environmental management problems. Many of the models available in the areas of mathematics, statistics, engineering, management, and social sciences have been found useful in solving an array of multi-faceted problems and are equally applicable in solving environmental problems. Specifically, papers that address optimization, or "satisficing" techniques to solve problems in the areas of environmental planning, resources allocation, biodiversity, and ecology, were of great interest to address some of the world's most pressing environmental problems. We successfully sought papers that may be conceptual, application-based, or theoretical.

The Special Issue presents five major papers that were published, each with uniqueness and contributions to the field of environmental planning and management. The views and models articulated in these articles can be beneficial in solving some of the environmental problems that confront the world today. This editorial summarizes the conclusions of our Special Issue which was highly successful. All the published papers have policy implications and are easier to read and interpret without losing the scientific component. It is our hope that the ideas espoused here will find wider applications in solving these important environmental problems. We aimed to address macro-environmental problems by understanding that environmental issues interface and interact with other subsystems on Earth. Environmental problems cannot be addressed by looking only at a problem as independent of all other subsystems that it interacts with. Some of the work has also attempted to draw parallels with sustainability development goals.

In this editorial, we shall briefly discuss the contributions of the five published papers in this Special Issue. We shall follow the sequence of publication by starting from the first published article to the last.

Shi, K., Zhou, Y., and Zhang, Z., in their article titled "Mapping the Research Trends of Household Waste Recycling: A Bibliometric Analysis", considered household waste recycling as a major cause of municipal solid waste pollution. They reviewed the status and mapped the research trend of research in household waste recycling published in the Web of Science database from 1991–2020. A bibliometrix analysis of these publications was carried out to identify the top contributing authors, countries, institutions, and journals. They note that most of the influential and well-cited articles focused on factors influencing residents' recycling behavior. Recycling behavior is viewed primarily from an sociopsychology and economics perspective. However, research in house waste recycling now includes other areas such as e-waste, source separation, life cycle assessment, sustainability, organic waste, and circular economy. These studies are increasingly becoming interdisciplinary, thus suggesting a systemic view of household waste recycling.

The article is very informative and shows an exponential growth rate in the number of publications on household waste recycling from 1991 to 2020. This growth rate signifies the growing interest in this research field and may very well align with the growing concerns of the general society about climate change and the resultant effects of pollution (see Figure 1 of the paper). China, the United Kingdom, and the USA seem to be the leading countries in terms of research, based on the number of citations of work in this area. This may also suggest growing environmental consciousness and awareness, and also represents the growing, and concerning, volume of waste in those countries. Interestingly, none of the developing countries appear to be on the top ten list and only China and Japan appear from the Asian countries. More awareness is required to solve household waste problems. Although the number of publications and citations may not be indicative of the environmental burden in a country, it may show the level of awareness and environmental consciousness. Then again, the database that was used may affect the number of publications that may be obtainable from the different countries. We understand the fact that the Web of Science is the leading database for quality research, but this may not necessarily cover some of the local publications that exist in many of the other countries. It is however imperative that we make all countries understand the consequences of household waste. Research collaborations with researchers from other countries that are not listed here may help to promote research interests and illuminate research works in household recycling.

Nnadi, V.E., Madu, C.N., and Ezeasor, I.C., in their article titled, "A systematic technic of prioritization of biodiversity conservation in Nigeria", developed a multi-criteria decision model to prioritize biodiversity conservation. This model is based on the use of analytic hierarchy process (AHP)—a multicriteria decision-making model. A group of biodiversity experts in the country was used to rank the three biodiversity conservation approaches, namely eco-system-, area-, and species-based approaches. The result showed the high performance of countries using eco-system-based approaches, followed by area-based and species-based approaches, respectively. The priorities developed were subsequently applied in resource allocation modeling. The research identifies areas for performance gap in the country and offers a policy-making approach for allocating limited resources to solve biodiversity conservation problems. This paper introduced management techniques and operational research models that can be used to address biodiversity problems. Although it is focused on a particular country, it has wider application since the frameworks and the model approaches presented can be applied in different settings.

Nantasaksiri, K., Charoen-amornkitt, P., Machimura, T., and Hayashi, K. titled their paper "Multi-Disciplinary Assessment of Napier Grass Plantation on Local Energetic, Environmental and Socioeconomic Industries: A Watershed-Scale Study in Southern Thailand". They investigated the potentials of using Napier grass in power generation in Southern Thailand. Napier grass is supposedly an energy crop that has far-reaching impacts on energy, environment, and socioeconomic features. A soil and water assessment tool (SWAT) is used to investigate its impacts on runoff, sediment, and nitrate loads in Songkhla Lake Basin (SLB), Southern Thailand. The results obtained show that Napier grass decreased the average surface runoff and sediment in the watershed. These results were applied in a multidisciplinary framework for decision support. It is shown that Napier grass provides benefits to hydrology and water quality when nitrogen fertilizers were applied at the levels of 0 and 125 kgN ha^{-1}. Conversely, benefits in terms of reducing energy supply, farmer's income, and carbon dioxide were highest when 500 kgN ha^{-1} of nitrogen fertilizer was applied. The paper concludes that the planting of Napier grass should be supported to increase energy supply; provide jobs; and reduce surface runoff, sediment yield, nitrate load, and carbon dioxide emission. The findings of this research are of particular importance, especially as we aim to decrease the demand on fossil fuel for energy consumption. Napier grass is a renewable source of energy. It is clean and sustainable and can contribute in reducing the generation of carbon dioxide, thereby helping to reduce global warming. While biogas energy generated from organic matters may release carbon in generating power, they are carbon-neutral, as crops also absorb the same amount of carbon during their growth. This is, however, not the only benefit of biogas as they also have higher yields and shorter life cycles. This ensures a stable fuel supply. This study should encourage an exploration of other organic matters and renewable resources, in terms of their efficiency in replacing, substituting, or reducing the use of fossil fuels.

H. Jiang and Y. He, in their paper titled "Evaluation of Optimal Policy on Environmental Change through Green Consumption", explored the association between green consumption and sustainable economic growth. They looked at the demand side of green consumption and how to use it to design an environmental policy package, in order to achieve economic growth and optimal social allocation. Using mathematical models, they concluded that: (1) green consumption does not necessarily have to be supply-side-driven to improve the environment; (2) green consumption driven by the demand side is better than the supply side in improving the environment and increasing the social welfare; (3) environmental change is more efficient when the environmental policy package includes green consumption. It is important to note that production activities that drive the economy will impact both the environment and social welfare. It is therefore significant to investigate the demand-side policy, which is exemplified by green consumption, and compare it to the supply-side policy, exemplified by green production on how they influence and/or are associated with environmental changes and social welfare.

Koo, J., Kim, J., Ryu, J., Shin, D.-S., Lee, S., Kim, M.-K., Jeong, J., and Lim, K.-J., in "Development of Novel QAPEX Analysis System Using Open-Source GIS", developed an Agricultural Policy/Environmental eXtender (APEX) interface that uses an open-source-based GIS software to simulate water quality impacts on various management practices for local farming activities. This model provides opportunities for farm/small watershed management and for local farmers, especially in developing countries, since there is no fee payment to use the interface. This new interface is also more flexible than the existing interface that requires paid license subscription. Furthermore, it can simulate hydrology and water quality with considerable precision. This model also presents visual output, making it easier to interpret simulation results. The open source model may also be used to derive data for sustainable agricultural practices and to develop effective policies on the different agricultural farming practices.

This Special Issue presents five different articles that took different approaches to address sustainability issues from planning and modeling perspectives. These articles are holistic in their considerations and are multidisciplinary as they adopt modeling approaches from other areas of learning, especially from management. They also emphasize the need to serve as decision support for policy and decision making. It is worth noting that to address the issues of sustainability and climate change, which are crucial in environmental planning and modeling, we must be cognizant of human involvement and the different worldviews that may inform such policies and decisions. The papers offer policy guidelines and framework, making the outcomes easier to implement. These articles address some of the complex issues in environmental planning and modeling. However, they are not exhaustive. Rather, they provide a stepping-stone for more work on developing approaches that can help to address these important problems. We do not necessarily need to start from scratch or reinvent the wheel. We can borrow from existing knowledge and models, and also take advantage of the multidisciplinary nature of environmental issues. We should also view the environment as a system that interacts with other systems and strive to develop a systemic approach to problem solving. The issues raised here are thought-provoking and aim to solve some crucial existing problems. The insights gained here could be used to solve other problems and to develop effective implementation plans. Ongoing research and future studies are required to continue to explore environmental problems and/or climate change issues from a holistic perspective, while considering other stakeholders to develop an efficient and effective solution.

Funding: This research received no external funding.

Conflicts of Interest: The author declared no conflict of interest.

 *sustainability*

Article

Evaluation of Optimal Policy on Environmental Change through Green Consumption

Haiwei Jiang [1] **and Yiyao He** [2,*]

[1] School of International Trade and Economics, Central University of Finance and Economics, Beijing 102206, China; hwjiang.work@gmail.com

[2] School of Economics, Hangzhou Normal University, Hangzhou 311121, China

*   Correspondence: heyiyao@zju.edu.cn

Abstract: This paper explores the relationship between green consumption and the environment from a new perspective of green consumption on the demand side. This paper further investigates how to design an environmental policy package to achieve optimal social allocation. The results show that: first, green consumption can still improve the environment without supply-driven policy; second, demand-driven environmental change is better than supply-driven change in improving the environment and increasing social welfare; and third, a policy package which includes green consumption is more efficient.

Keywords: green consumption; environmental change; environmental regulation; social welfare

1. Introduction

The relationship between economic growth and environmental change has always been an important issue in environmental economics. The environment provides exhaustible natural resources which are input into the production of nearly all goods, and thereby makes excellent contributions to sustainable economic growth [1,2]. Widespread industrialization and rapid economic development have seen the emergence of environmental problems—e.g., air pollution, global warming, and deforestation—which have influenced people's lives. In order to cope with these challenges, international organizations and many countries started to enact environmental policies to enhance environmental efficiency, increasing the sustainability of economic development [3]. An increasing number of people have shifted their purchase behavior towards being more environment-friendly and sustainable in the last few years. This paper categorizes environmental policies into demand-side policies (i.e., green consumption) and supply-side policies (i.e., green production), investigating how these two types of policies affect environmental change.

This paper develops a two-sector endogenous growth general equilibrium model (e.g., clean goods and non-clean goods) by incorporating environment evolvement. Following prior work, we study the relationship between economic growth and environmental change based on the heterogeneity of production, which can be clean or non-clean [4,5]. On the basis of market clearing conditions in the equilibrium, we conduct numerical analysis to quantitatively estimate the effects of green consumption and green production on environmental and social welfare. Furthermore, we show the path of optimal allocation by comparing impulse response between the decentralized equilibrium and social planner equilibrium. Finally, we analyze the mechanism of demand-side and supply-side environmental policies impacting the equilibrium, and illustrate the optimal policy package.

This paper relates to the literature on environmental regulation, green production, and green consumption. Economic growth will inevitably lead to environmental disasters without the government's environmental policies [4,6]. On the one hand, some studies argue that the government can mainly achieve environmental improvement by implementing supply-side environmental policies, such as environmental taxes, emission reduction

Citation: Jiang, H.; He, Y. Evaluation of Optimal Policy on Environmental Change through Green Consumption. *Sustainability* **2022**, *14*, 4869. https://doi.org/10.3390/su14094869

Academic Editor: Christian N. Madu

Received: 8 March 2022
Accepted: 16 April 2022
Published: 19 April 2022

Publisher's Note: MDPI stays neutral with regard to jurisdictional claims in published maps and institutional affiliations.

subsidies, emission charges [4,6,7]. These policies are able to secure a transition to green production, thereby slowing down the potential increases in pollution. On the other hand, the resource-intensive lifestyle of human beings has been regarded as one of the leading causes of environmental degradation [8]. Environmental policies aim to tackle environmental problems through encouraging consumers to adopt environmentally friendly behaviors, including using renewable energy and clean products [9]. Additionally, there is also some research studying the social welfare effects of environmental policies. These papers indicate that social welfare can be improved by implementing environmental subsidies or tax policies [10,11].

This paper also relates to the literature on green consumption and environmental change. The promotion of clean goods consumption can help to reduce some types of pollutants significantly [12]. For example, the utilization of renewable energy sources has a significant impact on environmental sustainability by decreasing air pollution [13]. Many studies find that green consumption improves the quality of the environment [14–18]. In addition, some research estimates the transformation path of consumption structure in different countries, and finds that the upward trend of transformation toward green consumption patterns significantly reduces pollution [19,20].

Compared with most existing macroeconomic literature which focuses on supply-side environmental regulation, this paper pays particular attention to demand-side environmental policies by profoundly studying the effect of green consumption on the environment, as well as the optimal package of governmental environmental policy to maximize social welfare. This paper shows that green consumption on the demand side can lead to environmental improvement. Furthermore, rather than enforcing supply-side environmental regulation alone, a policy package which includes green consumption is more efficient. This paper enriches the research on economic growth and environmental change, and provides meaningful policy implications for realizing sustainable economic development.

The rest of the paper is organized as follows. Section 2 constructs a two-sector endogenous growth general equilibrium model to study the decentralized equilibrium with green consumption. Section 3 is the numerical simulation. Section 4 is the extension, in which we further study the social planer equilibrium and the environmental policy package that can achieve the optimal social allocation. Section 5 concludes the paper.

2. Model: Decentralized Equilibrium

This section constructs a two-sector general equilibrium model. In the model, the representative household provides labor to obtain income and consumes clean and non-clean goods. The representative firms are completely competitive and can be divided into clean (c) and non-clean (d) parts where c is short for clean firm and d is short for dirty (non-clean) firm. They input capital and labor for production. Non-clean production has a negative impact on the environment. See Appendix A for detailed model construction and calculation process.

3. Quantitative Analysis

3.1. Calibration

There are three sets of calibrated parameters in Table 1. The first set is $\{\alpha, \beta, \eta, \delta, \zeta, \varepsilon\}$. Based on previous studies [21], set capital share $\alpha = 0.5$ which means that the output elasticity of capital is $1/2$, subjective discount factor $\beta = 0.995$ implying that the yearly interest rate is about 3%, inverse Frisch elasticity $\eta = 2$ which implies a Frisch elasticity of labor supply of 0.5, and depreciation rate $\delta = 0.025$ corresponds to an average 10% annual capital depreciation rate. Parameters ζ and ε are calibrated strictly following [4]'s method, which yields $\zeta = 2.46$ and $\varepsilon = 0.001$. This means that the initial temperature of simulation is defined as 1.31 °C, and when the temperature rises to 6 degrees centigrade, there will be an environmental disaster (for more details, please see [4]). The standard deviation (S.D.) of shocks, $\{\rho_{at}^{j}, \varepsilon_{at}^{j}, \rho_{\phi}, \varepsilon_{\phi t}\}$ are calibrated in the conventional sense. Persistence is equal to 0.9, and the standard deviation is set to 0.1, which means that the shock is moderate in this model, and the main results are not brought on by a strong exogenous shock. The third set is the steady-state of c_d/y_d. Among the eight categories of consumption expenditures published by the Chinese National Bureau of Statistics, four (food, clothing, household equipment supplies and maintenance services, and housing) belong to non-clean consumption, and their direct carbon emission indexes are higher than the overall average [20,22]. Therefore, based on the data on GDP per capita and per capita consumption expenditure of urban residents from the first quarter of 2005 to the fourth quarter of 2018, the ratio equals 36.45%. With the steady-state of c_d/y_d, this model can be better fitted into data in the following quantitative analysis.

Table 1. Parameter calibration.

Parameters	Value	Description
α	0.5	Capital share
β	0.995	Subjective discount factor
η	2	Inverse Frisch elasticity
δ	0.025	Depreciation rate
ζ	2.46	Environment disruption parameter
ε	0.001	Environment self-healing parameters
ρ_{at}^{j}	0.9	Persistence of TFP shock
ε_{at}^{j}	0.1	S.D. of TFP shock
ρ_ϕ	0.9	Persistence of clean consumption demand shock
$\varepsilon_{\phi t}$	0.1	S.D. of clean consumption demand shock
c_d/y_d	0.3645	Steady-state of non-clean consumption to non-clean output

Data source: Chinese National Bureau of Statistics.

3.2. Impulse Responses and Welfare Analysis

This part shows the impulse responses in Figure 1. According to Equations (A4) and (A5), when facing a positive clean-consumption demand shock, the household's demand for clean consumption increases rapidly, while the demand for non-clean consumption decreases relatively. Therefore, the former presents the positive impulse response, while the latter shows the negative. When the consumption structure transforms gradually to the clean side, both input and output of the non-clean (clean) firm decrease (increase). It is worth noting that according to Equation (A14), the impulse response of environment is positive, which means that the rising demand for clean products leads to environmental changes. In other words, even without the government's environmental policies, green consumption can still promote environmental improvement.

This paper compares "demand-driven" and "supply-driven" environmental changes, and shows the economic implications. Figures 2 and 3 show the respective impulse responses when the government executes a subsidy (tax) from supply side.

Figures 2 and 3 show that the subsidy (tax) to the clean (non-clean) firm increases labor input and output of the clean sector, while the non-clean sector decreases correspondingly. This result is caused by the government's environmental policies. Due to the production change, the household's demand for non-clean consumption decreases, while the demand for clean consumption increases. The environment improves. However, it is different from the demand-driven environmental change in Figure 1. First, when facing the same-size shock, the impulse responses of two types of consumption in Figure 1 is significantly larger than those in Figures 2 and 3. It means that the demand-driven force can directly influence the consumption goods market; in particular, the demand-driven force increases more clean demand and has a greater impact on the macroeconomy. Second, when facing the same-size shock, the impulse responses of the two types of outputs in Figure 1 are significantly larger than those in Figures 2 and 3. As proportional subsidy or tax distorts the relative price of two types of consumption, it correspondingly affects (distorts) outputs in supply-driven cases. Third, the environmental improvement in Figure 1 is significantly greater than those in Figures 2 and 3. Therefore, the result shows that demand-driven green consumption has a greater impact on economic fluctuations and environmental change; that is, the demand-driven environmental improvement performs better than the supply-driven.

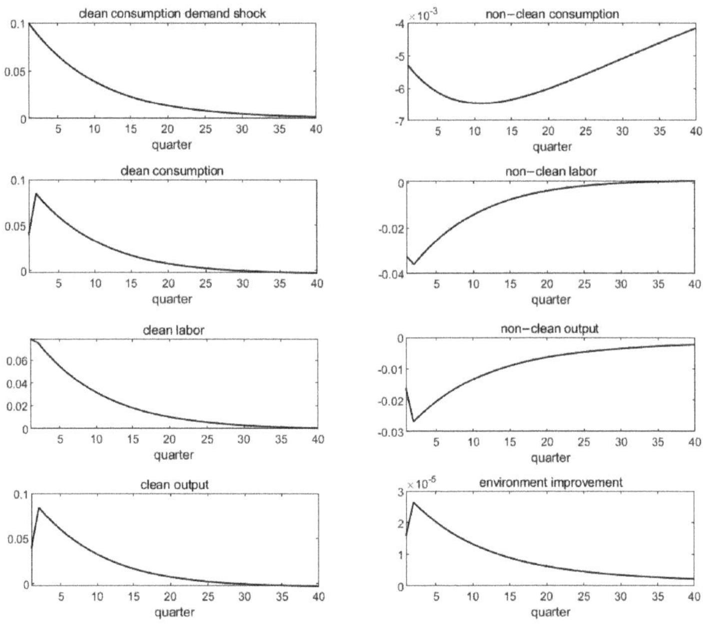

Figure 1. Impulse response (demand-driven).

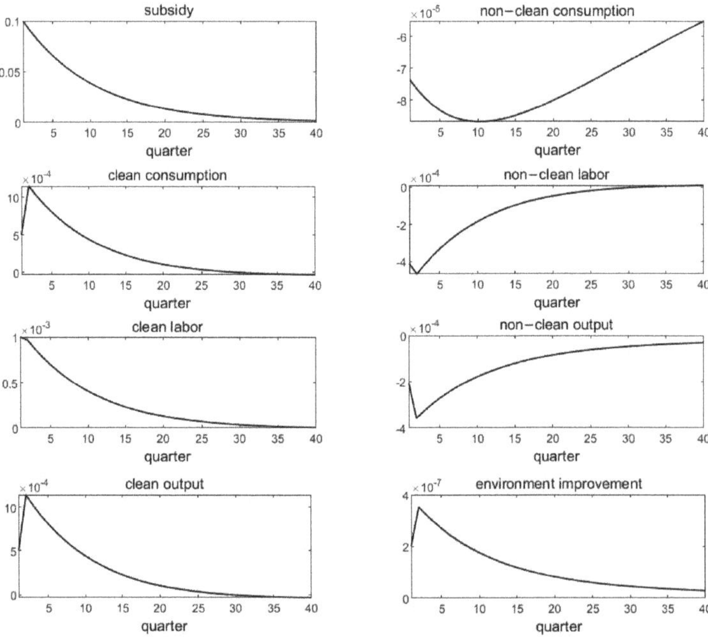

Figure 2. Impulse response (supply-driven, subsidy).

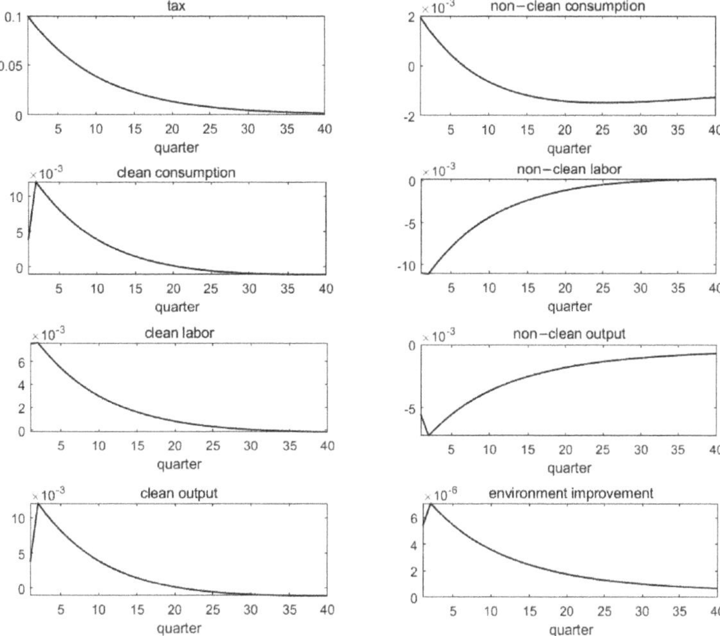

Figure 3. Impulse response (supply-driven, tax).

Furthermore, this paper studies the welfare effects of both demand-driven and supply-driven environmental change. The welfare under the benchmark policy regime is denoted by V_b; C^i_{dt}, C^i_{ct} and N^i_{ht} are the endogenous variables under different supply-driven cases. Welfare gains are calculated as the percentage decrease in consumption in perpetuity under each regime, such that the representative agent is indifferent between living under each regime [21]. The welfare gains are measured by Δ and satisfy $\log(1 - \Delta) = (V_b - V_a)/(1 - \beta)$, where V_a is the social welfare under any other case. Thus,

$$E\sum_{t=0}^{\infty} \beta^t \left[\log C^i_{dt}(1 - \Delta) + \phi_t \log C^i_{ct} - \varphi \frac{\left(N^i_{ht}\right)^{1+\eta}}{1+\eta}\right] = V_b$$

Table 2 shows that under both supply-driven cases, most variables fluctuate more, and social welfare decreases. When the government provides subsidy or tax to the firms, there is a distortion in the relative price of two types of consumption, which affects the firms' decisions on outputs and the household's consumption decisions. The distortion causes welfare losses. Therefore, combining the impulse responses and Table 2, this paper finds that demand-driven green consumption will lead to better environmental improvement (change) in both environmental change and social welfare. Even if there is no government policy intervention, economic growth will not lead to environmental disasters.

Table 2. Macroeconomic stability and welfare analysis.

	Demand-Driven	Supply-Driven (Subsidy)	Supply-Driven (Tax)
S.D. of non-clean consumption	0.0380	0.0429	0.0470
S.D. of clean consumption	0.1866	0.1807	0.1873
S.D. of labor supply	0.0212	0.0213	0.0210
S.D. of non-clean capital	0.0653	0.0703	0.0782
S.D. of clean capital	0.1978	0.1917	0.2009
Welfare gains (%)	-	-0.31×10^{-3}	-0.14×10^{-3}

Note: Social welfare in this paper is calculated by the Taylor first-order approximation of utility function [21].

4. Extension

4.1. Comparison of Two Equilibria

A natural question is whether the above decentralized equilibrium is the first best, and if not, what kind of environmental policy the government should adopt. Therefore, this paper solves the social planner problem where $j \in \{c, d\}$ in Appendix B.

This section reports the impulse response comparison between the decentralized and social planner equilibrium in Figure 4. First, when facing the same demand shock ϕ_t, clean consumption in social planner equilibrium increases more. Second, during the transformation of consumption structure, both the input and output of non-clean (clean) firm decrease (increase). Considering the negative impact of non-clean output on the environment from Equation (A21), non-clean output decreases more in the social planner equilibrium. Third, the impulse response of environment improvement is larger in the social planner equilibrium. It is consistent with the theoretical analysis above because social planners will take the negative environmental effects of non-clean output into account when allocating resources. Hence, Figures 1–4 show that the government can achieve the optimal allocation and accelerate environment improvement at the same time by environmental policy of green consumption. This paper uses log-linearization and only imposes a small standard deviation. Thus, the impulse response value is small, but this does not affect the main mechanism.

Table 3 further justifies that welfare in social planner equilibrium is higher. This result verifies the conclusion of the theory. This is because, according to the equation $\lambda_{dt} - \zeta\omega_t = \mu_{dt} = \hat{p}_{dt}$, the price ratio of both types of consumption goods expands from p_{ct}/p_{dt} to $p_{ct}/(p_{dt} - \zeta\omega_t)$ in social planner equilibrium. Meanwhile, the relative value of non-clean products becomes lower, which will inhibit the production of non-clean firm and contribute to the expansion of clean sector in the general equilibrium. Therefore, this paper argues that decentralized equilibrium is not the optimal social allocation. The government needs to regulate the economy in decentralized equilibrium, optimize the allocation of social resources, and improve the environment to the first best.

Table 3. Economic fluctuations and welfare analysis.

	Decentralized Equilibrium	Social Planner Equilibrium
S.D. of non-clean consumption	0.0380	0.0413
S.D. of clean consumption	0.1866	0.1983
S.D. of labor supply	0.0212	0.0173
S.D. of non-clean capital	0.0653	0.0834
S.D. of clean capital	0.1978	0.2145
Welfare gains (%)	-	6.21×10^{-4}

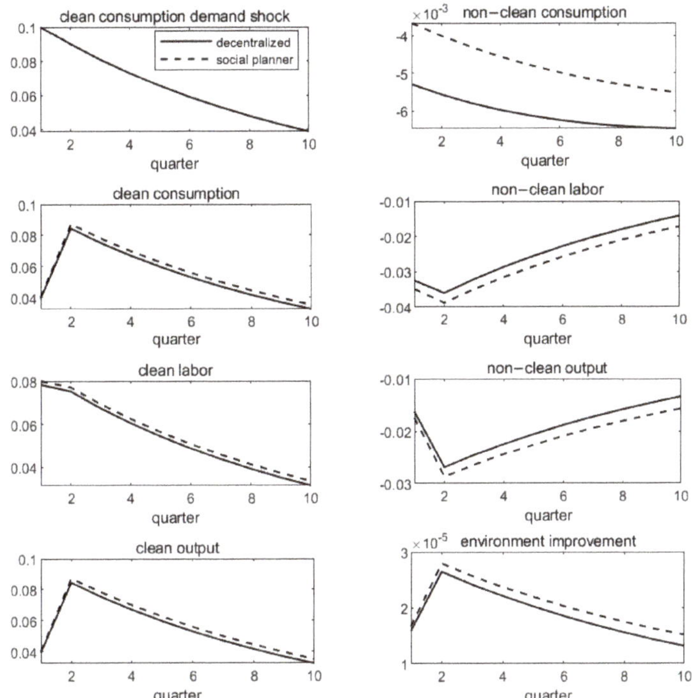

Figure 4. Impulse response (comparison).

4.2. Discussion on the Optimal Policy

The above social planner equilibrium depicts the optimal distribution. Corresponding with this reality, this paper will ask how the government can achieve the first best in the decentralized equilibrium through environmental policies. In order to answer this question, this paper must again clarify that the fundamental reason for the difference between decentralized equilibrium and social planner equilibrium lies in the different price ratio of clean and non-clean consumption goods. That is, the decentralized equilibrium can be equivalent to social planner equilibrium as long as p_{ct}/p_{dt} is expanded to $p_{ct}/(p_{dt} - \xi \omega_t)$. This means that considering the wedge of output to environment is the key to optimizing resource allocation in the decentralized equilibrium.

This section summarizes the environmental policy packages in Table 4, which can make the relative price $p_{ct}/p_{dt} \rightarrow p_{ct}/(p_{dt} - \xi \omega_t)$. It can be seen from Table 4 that environmental policy packages can be divided into two categories according to "whether it includes green consumption (demand side) or not". In case I-1, assume the government only subsidizes clean consumption by s_c^c to achieve the optimal allocation by moving the demand curve of clean goods outward to increase clean consumption, which yields $s_c^c = \frac{\xi \omega_t}{1 - \xi \omega_t}$. In case I-2, if the government subsidizes clean consumption and clean output by s_c^c and s_c respectively by moving the demand curve and supply curve of clean goods outward at the same time, then the two rate s_c^c and s_c satisfy the equation $(1 + s_c^c)(1 - s_c) = \frac{1}{1 - \xi \omega_t}$. In case I-3, if the government subsidizes clean consumption by s_c^c and levies environmental tax on non-clean output by τ_I, the demand curve of clean goods is moved outward and the supply curve of non-clean output is moved inward as tax τ_I decreases non-clean firm's production. In this case, we find that the two rates must follow $\frac{1 + s_c^c}{1 - \tau_I} = \frac{1}{1 - \xi \omega_t}$. Similarly, in case II-1, when the government only levies environmental tax on non-clean output by τ_{II}, the supply curve of non-clean output is pushed inward as τ_{II} suppresses non-clean output. By simple algebra, we get $\tau_{II} = \xi \omega_t$. In addition, if the government levies environmental tax on non-clean output by τ_{II} and subsidizes clean output by s_c, the two rates are in the relationship of $\frac{1-s_c}{1-\tau_{II}} = \frac{1}{1-\zeta\omega_t}$. The non-clean output supply curve moves inward to decrease output by tax τ_{II}, and at the same time, the clean output supply curve moves outward to increase production by subsidy s_c. If $s_c = 0$, it shows that "II-1" is a special case of "II-2".

In Table 4, the main result shows that the regulation efficiency of environmental policy containing green consumption is higher. Because "II-1" is a special case of "II-2 (when $s_c = 0$)", this paper can analyze the regulation efficiency of two types of policies by comparing "I-3" and "II-2" directly. Facing the same social optimal target, it is clear to see in Table 4 that $1 - \tau_I > 1 - \tau_{II} \Leftrightarrow \tau_I < \tau_{II}$. It means the tax rate with green consumption policy is lower, and the distortion of proportional tax on resource allocation is also smaller. Therefore, the policy regulation containing green consumption is more efficient, and the government should strengthen its support for green consumption. This result complements the literature which argues that government can mainly achieve environmental improvement by implementing supply-side environmental policies, such as environmental taxes, emission reduction subsidies, and emission charges [4,6,7]. This paper suggests demand-side policy which emphasizes green consumption can achieve higher efficiency and welfare gains.

We also find that as the environmental wedge $\zeta\omega_t$ rises, subsidies for green consumption should also increase. In the case of "I-1", $s_c^c = \zeta\omega_t/(1 - \zeta\omega_t)$ is a monotonic increasing function of $\zeta\omega_t$. That is, when the cost of environmental damage $\zeta\omega_t$ increases, the subsidy for green consumption should also be strengthened to achieve optimal social distribution. In addition, in the cases of "I-2" (s_c^c, s_c) and "I-3" (s_c^c, τ_I), there is a scientific trade-off between policies of demand-side and supply-side. Specifically, there is a significant positive correlation between subsidizing clean consumption s_c^c and subsidizing clean output s_c, and there is a significant negative correlation between subsidizing green consumption s_c^c and levying environmental tax τ_I on non-clean output. This is because the greater the subsidy for clean consumption s_c^c, the higher the household demand for clean consumption goods and output in the general equilibrium. Therefore, the subsidy for clean output s_c will also increase. While the subsidy for clean consumption s_c^c is greater, the environmental regulation on the demand side can well raise the relative price p_{ct}/p_{dt}. At this time, the punishment of the supply side for non-clean output τ_I can be appropriately relaxed, and the social distribution can still be optimal. These results extend the supply-side literature on environmental change, and enrich the policy implication of environmental regulation. The government has many options to regulate environment when considering the trade-off between green consumption (demand-side) and environmental taxes (supply-side) instead of being limited to only supply-side policies as most macroeconomic papers did.

Table 4. The optimal environmental policy package.

	Policy	Value	Mechanism
I: with green consumption	I-1: only subsidize clean consumption by s_c^c	$s_c^c = \frac{\zeta\omega_t}{1-\zeta\omega_t}$	The demand curve of clean goods moves outward to increase its consumption.
	I-2: subsidize clean consumption by s_c^c+ subsidize clean output by s_c	$(1 + s_c^c)(1 - s_c) = \frac{1}{1-\zeta\omega_t}$	The demand curve and supply curve of clean goods move outward at the same time.
	I-3: subsidize clean consumption by s_c^c+ environmental tax on non-clean output by τ_I	$\frac{1+s_c^c}{1-\tau_I} = \frac{1}{1-\zeta\omega_t}$	The demand curve of clean goods moves outward and the supply curve of non-clean output moves inward.
II: no green consumption	II-1: only environmental tax on non-clean output by τ_{II}	$\tau_{II} = \zeta\omega_t$	The supply curve of non-clean output moves inward to suppress its output incentive.
	II-2: environmental tax on non-clean output by τ_{II}+ subsidize clean output by s_c	$\frac{1-s_c}{1-\tau_{II}} = \frac{1}{1-\zeta\omega_t}$	The non-clean output supply curve moves inward to reduce the output, and at the same time moves the clean output supply curve outward to stimulate production.

Note: If only the clean output is subsidized, the relative price p_{ct}/p_{dt} will decrease by moving the clean output supply curve outward with other conditions unchanged. Although this policy can increase clean output and improve environmental quality, the result is not the first best.

4.3. Discussion of the Results

With the improvement of living standards, people's awareness of environmental protection is strengthened. They are increasingly interested in green products and are keen on green consumption. Green production and developing green markets have become new trends in the 21st century.

For example, the Chinese government has been encouraging green consumption, subsidizing green production and levying taxes on non-clean output. The government attaches great importance to promoting new energy vehicles and has successively issued a series of policies, including financial subsidies, exemption from vehicle purchase tax, and increasing loans, which have promoted the consumption of new energy vehicles. Moreover, the government imposes punitive taxes on air pollutants, water pollutants, solid waste emissions, and noise [23]. With the reform of China's urban heating system, the development and utilization of new energy, the adjustment of industrial structure, and the return of farmland to forests, China's carbon emissions have decreased significantly [24–26]. As China's regional development is unbalanced, policies should be adjusted to local conditions [27,28].

On the demand side, the term "green consumers" refers to those consumers who care about the ecological environment and have purchase intention for green products. They have green consciousness and have or may transform green consciousness into green consumption behavior. The Chinese government subsidizes green consumption directly as promotion. On the supply side, there were 50 categories of 200 million green basket commodities on Alibaba's online retail platform in 2015 (the term "green basket commodities" refers to the collection of commodities with three green attributes of "capital saving and energy saving, environment-friendly, and health quality"), most of which are subsidized by the government. The consumption of green basket accounts for 11.5% of Alibaba's retail platform, and the compound annual growth rate over the past five years has exceeded 80%.

By analyzing the shopping behavior of 400 million consumers on the Alibaba China retail platform, the Alibaba Research Institute found that the online population in line with the characteristics of green consumers reached 65 million, accounting for 16% of the active users of Taobao, an increase of 14 times in the last four years. The release of green consumer demand is bound to better guide green supply and promote supply side reform. Besides, according to the analysis of the consumption frequency of green basket commodities, in the past five years, heavy green consumers (with an average annual consumption of more than 20 times) have significantly expanded, increasing from 19.4% in 2011 to 28.4% in 2015.

In 2015, the Alibaba online retail platform reduced the emission of about 30 million tons of carbon dioxide by saving energy and material consumption, which is equivalent to adding forests the size of Poyang Lake in China. The water-saving products sold on the platform can save water for 13 days in Beijing every year. The annual power saving of energy-saving products can be used for 25 days in Beijing. The environmentally friendly packaging products sold on the platform can reduce the consumption of plastic bags and convert them into oil, which can be used by Beijing taxis for 62 days. In 2016, the comprehensive utilization of waste textiles in China was 3.6 million tons, which could save 4.6 million tons of crude oil and 270,000 hectares of cultivated land. The energy-efficient air conditioners, refrigerators, washing machines, and water heaters sold in China in 2017 can save about 10 billion kwh of electricity annually, which is equivalent to reducing 6.5 million tons of carbon dioxide, 14,000 tons of sulfur dioxide, 14,000 tons of nitrogen oxides, and 11,000 tons of particulate matter. All of these actions promote the environment.

5. Conclusions

The relationship between the environment and the macroeconomy is always a heated issue. The literature mainly focuses on supply-side environmental regulation, and argues that economic growth will not cause environmental disasters only if the government implements supply-side environmental policies. However, this paper constructs a two-sector endogenous growth general equilibrium model to study the relationship between green consumption and the environment from a new demand-driven perspective. We find that green consumption can promote environmental improvement without the government's intervention. Besides, demand-driven environmental change is better than the supply-driven in improving environmental change and social welfare.

Then this paper studies the relationship between environmental regulation and the environment in a social planer model. The results show that the welfare of decentralized equilibrium is about 0.0006% lower at the Taylor first order approximation. Moreover, policymakers need to optimize the allocation of social resources and improve the environment to the first best. We show that the policy package which includes green consumption is more efficient. Specifically, the demand-side policy emphasizing green consumption can achieve higher efficiency and welfare gains. Moreover, the government has many options to regulate environment when considering the trade-off between green consumption (demand-side) and environmental taxes (supply-side). For example, the government can subsidize green consumption increasing with environmental disruption and clean output, or decreasing with environmental tax of non-clean output.

To promote sustainable economic development, policymakers should not only pay attention to the punishment of environmental damage, but also strengthen the support for green consumption. Overall, this paper argues for a multi-level environmental regulation system of "green consumption and green production", especially the support for green consumption to promote sustainable economic development.

As this paper is most related to the theoretical framework of macroeconomy, we follow the approach of this strand of literature [4,6,7] and assume that the representative household is homothetic. They need to consume both clean goods that are environmental-friendly (e.g., electric vehicles) and non-clean goods that are contaminative (e.g., paper or coal). Because the household's utility is an increasing function of clean goods, all the homothetic households in this model will agree to protect the environment by consuming more clean goods. As a result, social welfare gains due to less resource misallocation between clean and non-clean sectors and increasing utility, besides the decreasing distortion of proportional tax. It is verified by many empirical papers [19,20] that the upward trend of transformation toward green consumption patterns significantly reduces pollution, and the trend is more pronounced especially in developing countries. In this model, any awareness program being considered for citizens to understand the green consumption concept is included in the demand shock of clean consumption. It implies that if a program proposes a green consumption concept, the household will be more willing to increase clean consumption with a positive clean-consumption demand shock. We think that considering the heterogeneous household preferences for green consumption in the model will be very meaningful in the future. If we introduce heterogeneous household preferences, people at all levels of society may prefer green consumption, as it reduces resource misallocation and raises utility and social welfare at the same time. Interestingly, the government may need to provide higher subsidy for the low-income group to encourage their green consumption.

This paper does not incorporate the associated energy use of green manufacturing and consumption. In the model, this phenomenon is described by the pollution parameter of non-clean firm ζ. Because different industries have different production habits and impacts on the environment, the relative policy, such as subsidy and tax, can be differentiated. Therefore, research discussing heterogeneous industries is also a feasible direction in the future. This paper does not consider the impact of an open economy on green consumption and environmental change. As the carbon emission rights can be traded across borders, the exchange rate and capital flow can affect domestic green consumption and environmental change through the cross-border transaction of carbon emission rights in an open economy; this is worth carefully studying in the future.

Author Contributions: Writing—original draft, Y.H.; Writing—review & editing, H.J. All authors have read and agreed to the published version of the manuscript.

Funding: This research was funded by the National Natural Science Foundation of China (grant number 71904169) and the Scientific Research Foundation for Scholars of HZNU (grant number 4015C50221204101).

Institutional Review Board Statement: Not applicable.

Informed Consent Statement: Not applicable.

Data Availability Statement: Not applicable.

Conflicts of Interest: The authors declare no conflict of interest.

Appendix A. Decentralized Equilibrium

A.1. The Representative Household

The household has the utility function

$$E_t \sum_{t=0}^{\infty} \beta^t \left[\log(C_{dt}) + \phi_t \log(C_{ct}) - \varphi \frac{N_{ht}^{1+\eta}}{1+\eta} \right] \tag{A1}$$

where the utility weight parameter $\varphi > 0$ and the subjective discount factor $\beta \in (0,1)$, C_{ct} and C_{dt} denote the clean and non-clean consumption at t respectively, N_{ht} denotes labor supply. Assume ϕ_t as the demand shock of clean consumption which follows the AR(1) process [29]

$$\log \phi_t = \rho_\phi \log \phi_{t-1} + \varepsilon_{\phi t} \tag{A2}$$

The parameter $\rho_\phi \in (-1,1)$, and $\varepsilon_{\phi t}$ is i.i.d. standard normal processes.

Normalize the price of non-clean consumption to 1, denote W_t is the household income, π_t is the dividend, then the household's budget constraint is given by

$$C_{dt} + P_{ct} C_{ct} = W_t N_{ht} + \pi_t \tag{A3}$$

The household chooses $\{C_{ct}, C_{dt}, N_{ht}\}$ to maximize Equation (A1) subject to Equations (A2) and (A3). The first order conditions are

$$1/C_{dt} = \lambda_t \tag{A4}$$

$$\phi_t/C_{ct} = \lambda_t P_{ct} \tag{A5}$$

$$\varphi N_{ht}^{\eta} = \lambda_t W_t \tag{A6}$$

where λ_t is the Lagrange multiplier of Equation (A3). Equation (A6) shows that wage is equal to the marginal substitution rate of consumption and leisure, and household purchases more clean and non-clean goods with increasing income.

A.2. The Representative Firms

Clean and non-clean firms are perfectly competitive. Firms input labor N_{jt} and capital K_{jt} to produce ($j \in \{c,d\}$), and denote A_{jt} is the total factor productivity (TFP). The production function is

$$Y_{jt} = A_{jt} K_{jt}^{\alpha} N_{jt}^{1-\alpha} \tag{A7}$$

where α is the capital share and A_{jt} follows the AR(1) stochastic process

$$\log A_{jt} = \rho_a^j \log A_{j,t-1} + \varepsilon_{at}^j \tag{A8}$$

The parameter $\rho_a^j \in (-1,1)$, and ε_{at}^j is i.i.d. standard normal processes. The capital accumulation satisfies

$$K_{j,t+1} = (1-\delta)K_{jt} + I_{jt} \tag{A9}$$

where δ is the capital depreciation rate. The firm chooses the optimal $\{K_{j,t+1}, N_{jt}\}(j \in \{c,d\})$ to maximize profit subject to Equations (A7)–(A9)

$$\max_{\{K_{j,t+1}, N_{jt}\}} E_t \sum_{t=0}^{\infty} \beta^t \frac{\lambda_t}{\lambda_0} \left\{ P_{jt} A_{jt} K_{jt}^{\alpha} N_{jt}^{1-\alpha} - W_t N_{jt} - I_{jt} \right\}$$

and the first order conditions are

$$1 = \beta E_t \lambda_{t+1} / \lambda_t (\alpha P_{j,t+1} Y_{j,t+1} / K_{j,t+1} + 1 - \delta) \tag{A10}$$

$$(1 - \alpha) P_{jt} Y_{jt} = W_t N_{jt} \tag{A11}$$

Equation (A10) is the Euler equation of capital, which means that the price of capital is equal to the discounted present value of its future marginal products plus the undepreciated capital. Equation (A11) is the labor demand function, which implies that the real wage is equal to the marginal product of labor. Both equations mean that in the absence of environmental policies, non-clean firm will not consider the negative output externality to the environment.

A.3. Market Clearing Conditions and Equilibrium

In a competitive equilibrium, the goods market clearing conditions imply that

$$C_{ct} = Y_{ct} \tag{A12}$$

$$C_{dt} + I_t = Y_{dt} \tag{A13}$$

Environment S_{t+1} evolves as

$$S_{t+1} = -\zeta Y_{dt} + (1 + \varepsilon) S_t \tag{A14}$$

where $\zeta > 0$ is the pollution parameter of non-clean firm, ε is the recovery parameter of environment, $S_t \in (0, \bar{S})$. The labor market clearing condition implies that

$$N_{ct} + N_{dt} = N_{ht} \tag{A15}$$

A competitive equilibrium consists of sequences of endogenous variables $\{C_{ct}, C_{dt}, N_{ht}, N_{ct}, N_{dt}, I_{ct}, I_{dt}, K_{ct}, K_{dt}, Y_{ct}, Y_{dt}, S_{t+1}\}_{t=0}^{\infty}$, such that (i) taking the prices $\{P_{ct}, W_t\}$ as given, the allocations solve the optimizing problems for the household and the firms, and (ii) all markets clear.

Appendix B. Social Planner Equilibrium

This paper also solves the social planner problem where $j \in \{c,d\}$.

$$E_t \sum_{t=0}^{\infty} \beta^t \left[\log(C_{dt}) + \phi_t \log(C_{ct}) - \varphi \frac{N_{ht}^{1+\eta}}{1+\eta} \right],$$

$$s.t. C_{ct} = Y_{ct} \tag{A16}$$

$$C_{dt} + [K_{d,t+1} - (1 - \delta) K_{dt} + K_{c,t+1} - (1 - \delta) K_{ct}] = Y_{dt} \tag{A17}$$

$$Y_{ct} = A_{ct} K_{ct}^{\alpha} N_{ct}^{1-\alpha} \tag{A18}$$

$$Y_{dt} = A_{dt} K_{dt}^{\alpha} N_{dt}^{1-\alpha} \tag{A19}$$

$$S_{t+1} = -\zeta Y_{dt} + (1 + \varepsilon) S_t \tag{A20}$$

where $N_{ct} + N_{dt} = N_{ht}$, $\log A_{jt} = \rho_a^j \log A_{j,t-1} + \varepsilon_{at}^j$ $\log A_{jt} = \rho_a^j \log A_{j,t-1} + \varepsilon_{at}^j$ and $\log \phi_t = \rho_\phi \log \phi_{t-1} + \varepsilon_{\phi t}$.

Denote λ_{ct}, λ_{dt}, μ_{ct}, μ_{dt} and ω_t are the Lagrange multipliers of Equations (A16)–(A20) respectively, it is easy to get the optimal distributions are

$$\text{FOC of } Y_{ct} \text{ and } Y_{dt}: \ \lambda_{ct} = \mu_{ct} = \hat{p}_{ct}, \ \lambda_{dt} - \zeta\omega_t = \mu_{dt} = \hat{p}_{dt}, \tag{A21}$$

$$\text{FOC of } C_{ct} \text{ and } C_{dt}: \ \phi_t/C_{ct} = \lambda_{ct}, \ 1/C_{dt} = \lambda_{dt}, \tag{A22}$$

$$\text{FOC of } N_{ct} \text{ and } N_{dt}: \ \varphi N_{ht}^{\eta} = (1-\alpha)\mu_{jt}Y_{jt}/N_{jt}, \tag{A23}$$

$$\text{FOC of } K_{c,t+1} \text{ and } K_{d,t+1}: \ \lambda_{dt} = \beta E_t\left[\lambda_{d,t+1}(1-\delta) + \alpha\mu_{j,t+1}Y_{j,t+1}/K_{j,t+1}\right], \tag{A24}$$

$$\text{FOC of } S_t: \ \omega_t = \beta E_t\omega_{t+1}(1+\varepsilon). \tag{A25}$$

Equation (A21) is the most important equation in this paper, where $\mu_{ct} = \hat{p}_{ct}$ and $\mu_{dt} = \hat{p}_{dt}$ respectively represent the shadow price of two sectors in the first best equilibrium. Because clean production has no destructive effect on the environment, the marginal price of clean product is equal to its marginal utility. However, in social planner equilibrium, there is one more $\zeta\omega_t$ in the first order condition of Y_{dt}, which means the marginal damage cost to environment for each additional unit of non-clean output (every one more unit Y_{dt} will damage ζ environment in the next period). This implies that by introducing a wedge $\zeta\omega_t$, the government makes the marginal price of non-clean product equal to its marginal utility minus environmental cost. Equation (A21) shows that the environmental wedge decreases $\hat{p}_{dt}$, thus reducing the non-clean output. This is a different result from the decentralized equilibrium in which $\lambda_{dt} = \mu_{dt} = P_{dt}$. Because the non-clean sector only cares about the profit maximization within the enterprise and does not consider the damage of its output to environment, the marginal price of non-clean product is higher than the social optimal; that is, the non-clean sector will not impose environmental regulation on itself. Equation (A22) is the Euler equation of consumption, which means that increasing environmental wedge will promote the increase of the relative demand for clean product at a given price. Equations (A23) and (A24) are the standard Euler equations of production factors. The marginal outputs of labor and capital are equal to their respective shadow prices. Equation (A25) indicates that the shadow price of environmental quality in the period t is equal to its discounted present value of the shadow price in the period $t+1$ after recovering $(1+\varepsilon)$. From Equations (A21)–(A25), it can be seen that the decentralized equilibrium obviously does not reach social optimization. The fundamental reason is that in the decentralized equilibrium, the production of non-clean firm does not consider its environmental pollution cost, while the social planner cares about this cost in the equilibrium. Therefore, in the two kinds of equilibria, the relative prices of clean and non-clean product are different, and the relative price p_{ct}/p_{dt} in decentralized equilibrium is distorted. This paper will use numerical simulation in the next section to further demonstrate this theoretical conclusion.

References

1. Grossman, G.M.; Krueger, A.B. Economic growth and the environment. *Q. J. Econ.* **1995**, *110*, 353–377. [CrossRef]
2. Wu, H.; Hao, Y.; Ren, S. How do environmental regulation and environmental decentralization affect green total factor energy efficiency: Evidence from China. *Energy Econ.* **2020**, *91*, 104880. [CrossRef]
3. Cardenas, L.M.; Franco, C.J.; Dyner, I. Assessing emissions–mitigation energy policy under integrated supply and demand analysis: The Colombian case. *J. Clean. Prod.* **2016**, *112*, 3759–3773. [CrossRef]
4. Acemoglu, D.; Aghion, P.; Bursztyn, L.; Hemous, D. The environment and directed technical change. *Am. Econ. Rev.* **2012**, *102*, 131–166. [CrossRef]
5. Grimaud, A.; Rouge, L. Environment, directed technical change and economic policy. *Environ. Resour. Econ.* **2008**, *41*, 439–463. [CrossRef]
6. Acemoglu, D.; Akcigit, U.; Hanley, D.; Kerr, W. Transition to clean technology. *J. Political Econ.* **2016**, *124*, 52–104. [CrossRef]
7. Acemoglu, D.; Rafey, W. *Mirage on the Horizon: Geo-Engineering and Carbon Taxation without Commitment*; Working Paper No.24411; National Bureau of Economic Research: Cambridge, MA, USA, 2018.
8. Tripathi, A.; Singh, M.P. Determinants of sustainable/green consumption: A review. *Int. J. Environ. Technol. Manag.* **2016**, *19*, 316–358. [CrossRef]

9. Testa, F.; Pretner, G.; Iovino, R.; Bianchi, G.; Tessitore, S.; Iraldo, F. Drivers to green consumption: A systematic review. *Environ. Dev. Sustain.* **2021**, *23*, 4826–4880. [CrossRef]
10. Wang, L.; Lu, J. Analysis of the social welfare effect of environmental regulation policy based on a market structure perspective and consumer. *Sustainability* **2019**, *12*, 104. [CrossRef]
11. Kotani, K.; Kakinaka, M. Some implications of environmental regulation on social welfare under learning-by-doing of eco-products. *Environ. Econ. Policy Stud.* **2017**, *19*, 121–149. [CrossRef]
12. Kahia, M.; Aïssa, M.S.B.; Lanouar, C. Renewable and nonrenewable energy use-economic growth nexus: The case of MENA net oil importing countries. *Renew. Sustain. Energy Rev.* **2017**, *71*, 127–140. [CrossRef]
13. Bhattacharya, M.; Churchill, S.A.; Paramati, S.R. The dynamic impact of renewable energy and institutions on economic output and CO_2 emissions across regions. *Renew. Energy* **2017**, *111*, 157–167. [CrossRef]
14. Jaforullah, M.; King, A. Does the use of renewable energy sources mitigate CO_2 emissions? A reassessment of the US evidence. *Energy Econ.* **2015**, *49*, 711–717. [CrossRef]
15. Salahuddin, M.; Gow, J.; Ozturk, I. Is the long-run relationship between economic growth, electricity consumption, carbon dioxide emissions and financial development in Gulf Cooperation Council Countries robust? *Renew. Sustain. Energy Rev.* **2015**, *51*, 317–326. [CrossRef]
16. Saidi, K.; Omri, A. Reducing CO_2 emissions in OECD countries: Do renewable and nuclear energy matter? *Prog. Nucl. Energy* **2020**, *126*, 103425. [CrossRef]
17. Jebli, M.B.; Farhani, S.; Guesmi, K. Renewable energy, CO_2 emissions and value added: Empirical evidence from countries with different income levels. *Structural Change Econ. Dyn.* **2020**, *53*, 402–410. [CrossRef]
18. Kirikkaleli, D.; Adebayo, T.S. Do renewable energy consumption and financial development matter for environmental sustainability? New global evidence. *Sustain. Dev.* **2021**, *29*, 583–594. [CrossRef]
19. Feng, T.; Sun, L.; Zhang, Y. The relationship between energy consumption structure, economic structure and energy intensity in China. *Energy Policy* **2009**, *37*, 5475–5483. [CrossRef]
20. Yuan, B.; Ren, S.; Chen, X. The effects of urbanization, consumption ratio and consumption structure on residential indirect CO2 emissions in China: A regional comparative analysis. *Appl. Energy* **2015**, *140*, 94–106. [CrossRef]
21. Chang, C.; Liu, Z.; Spiegel, M.M. Capital controls and optimal Chinese monetary policy. *J. Monet. Econ.* **2015**, *74*, 1–15. [CrossRef]
22. Wei, Y.; Liu, L.; Fan, Y.; Wu, G. The impact of lifestyle on energy use and CO_2 emission: An empirical analysis of China's residents. *Energy Policy* **2007**, *35*, 247–257. [CrossRef]
23. John, A.; Pecchenino, R. An overlapping generations model of growth and the environment. *Econ. J.* **1994**, *104*, 1393–1410. [CrossRef]
24. Dasgupta, S.; Laplante, B.; Wang, H.; Wheeler, D. Confronting the Environmental Kuznets Curve. *J. Econ. Perspect.* **2002**, *16*, 147–168. [CrossRef]
25. Song, T.; Zheng, T.; Tong, L. An empirical test of the Environmental Kuznets Curve in China: A panel cointegration approach. *China Econ. Rev.* **2008**, *19*, 381–392. [CrossRef]
26. Gui, S.; Zhao, L.; Zhang, Z. Does municipal solid waste generation in China support the Environmental Kuznets Curve? New evidence from spatial linkage analysis. *Waste Manag.* **2019**, *84*, 310–319. [CrossRef]
27. Shen, J. A simultaneous estimation of Environmental Kuznets Curve: Evidence from China. *China Econ. Rev.* **2006**, *17*, 383–394. [CrossRef]
28. Song, M.; Zhang, W.; Wang, S. Inflection point of Environmental Kuznets Curve in mainland China. *Energy Policy* **2013**, *57*, 14–20. [CrossRef]
29. Liu, Z.; Wang, P.; Zha, T. Land-price dynamics and macroeconomic fluctuation. *Econometrica* **2013**, *8*, 1147–1184.

 *sustainability*

Article

Mapping the Research Trends of Household Waste Recycling: A Bibliometric Analysis

Kun Shi, Yi Zhou and Zhen Zhang *

Department of Environmental Science & Engineering, Fudan University, Shanghai 200433, China;
18210740035@fudan.edu.cn (K.S.); 18210740074@fudan.edu.cn (Y.Z.)
* Correspondence: zhenzhang@fudan.edu.cn

Abstract: Household waste recycling has been widely considered the key to reducing the pollution caused by municipal solid waste and promoting sustainable development. This article aims to clarify the status and map the research trends in the field of household waste recycling. Bibliometric analysis is performed using bibliometrix based on publications during 1991–2020 in the Web of Science database. Results show that academic output in this field is growing rapidly. The top contributing authors, countries, institutions, and journals are identified. Collaboration network of authors, institutions, and countries are created and visualized. The most influential and cited articles in this field mainly focus on factors influencing residents' recycling behavior from the perspectives of sociopsychology and economics. The theory of planned behavior is the most widely used psychological model. Other research hotspots include electronic waste, source separation, life cycle assessment, sustainability, organic waste, and circular economy. Studies on household waste recycling have become more and more comprehensive and interdisciplinary with the evolution of research themes.

Keywords: bibliometric analysis; bibliometrix; household recycling; research trends; science mapping; waste management

Citation: Shi, K.; Zhou, Y.; Zhang, Z. Mapping the Research Trends of Household Waste Recycling: A Bibliometric Analysis. *Sustainability* **2021**, *13*, 6029. https://doi.org/10.3390/su13116029

Academic Editor: Christian N. Madu

Received: 12 May 2021
Accepted: 26 May 2021
Published: 27 May 2021

Publisher's Note: MDPI stays neutral with regard to jurisdictional claims in published maps and institutional affiliations.

1. Introduction

Currently, municipal solid waste management has become one of the most critical environmental issues in the world [1]. Environmental pollution caused by municipal solid waste has led to health threats to residents [2]. With the continuous growth of world's population and the improvement of industrialization and urbanization, the problem of waste management is expected to become more serious in the future [3]. The continuous increase in the amount of municipal solid waste restricts the improvement of residents' quality of life and has become an important obstacle to sustainable development [4].

Municipal solid waste includes but is not limited to household waste, industrial waste, commercial waste, construction and demolition waste, and waste generated from schools, hospitals, and road sweeping [5]. The majority of municipal solid waste is contributed to by household waste [6]. Compared with other sources of waste, the composition of household waste is more complex [7], which results in greater challenges for urban management. In this context, promoting the recycling of household waste is critical to reduce the volumes of waste generated, conserve natural resources, and move towards a circular economy [8].

Given that household waste recycling has become an area of increasing concern, and related recycling schemes have been put into practice in many countries [9–13], the amount of academic research on certain topics has increased significantly in recent years. These studies focus on various aspects such as environmental science and technology, sociology, psychology, and economics. Specifically, the majority of the studies focus on factors influencing residents' attitudes toward recycling and determinants of their recycling behavior, as the success of household waste recycling programs depends on residents' consistently positive participation and their support for recycling policies [14]. Psychological factors including social norms [15], moral norms [16], environmental concern [17], recycling habits [18], and past experience [19] have been examined and proven to have significant influence on individual's recycling behavior and willingness to recycle. Other studies have emphasized the importance of objective situational factors such as the convenience of recycling [20] and laws and regulations [21,22]. From an economic perspective, people tend not to participate in recycling when the expected utility of recycling is not enough to cover the cost [8]. Hence, the introduction of financial incentives could be an effective driver for promoting recycling programs [23].

However, with the rapid increase in the number of academic publications year by year, it is becoming increasingly difficult to keep up with the all the latest studies. Voluminous and fragmented research streams hamper the accumulation of knowledge and empirical evidences from previous research papers. In this case, researchers would spend excessive time and effort when reviewing the literature, and they may still not be able to get a full picture of a certain research field. Therefore, bibliometric methods have been developed for structured analysis of large amounts of literature information and identification of the research trends of an area based on the statistical measurement of scientific productions [24]. Several studies have already applied bibliometric analysis to the field of waste management. Li et al. (2018) [25] analyzed the research trends on solid waste reuse and recycling from 1992 to 2016. Some researchers focused on specific types of municipal solid waste, such as construction and demolition waste [26,27] and electronic waste [28,29]. Tsai et al. (2020) [30] performed bibliometric analysis on municipal solid waste management in the context of the circular economy. Wang et al. (2020) [31] combined bibliometrics with text-mining and reviewed the evolution of municipal waste management. According to these existing bibliometric studies, recycling is one of the key issues of waste management. Furthermore, Tsai et al. [30] emphasized that the promotion and enhancement of household waste recycling programs is a future challenge and direction, so it is essential to understand personal recycling behaviors. Meanwhile, there is still lack of bibliometric studies focusing on household waste recycling, as no such publications have been found.

The present study aims to map the research trends of household waste recycling based on bibliometric approaches. Specifically, the objectives of this article are to answer the following questions:

1. Which authors, journals, institutions, and countries have contributed the most to this field of research?
2. How do the authors, institutions, and countries relate to and cooperate with each other in academic research?
3. What are the most influential articles that are worth attention, and what are their arguments?
4. What are the research hotspots in this field, and how do they evolve over time?

By filling the gaps in bibliometric analysis in the field of household recycling, it is hoped that this article can provide a "big picture" of the field for future studies and help researchers find topics worth studying. In the next section, the data acquisition process and the methodology are explained in detail. Section 3 presents the results of the bibliometric analysis, including an overview of the scientific production, collaboration network analysis, citation analysis, and conceptual structure analysis. Section 4 presents the conclusions and prospects for future research.

2. Materials and Methods

2.1. General Workflow

This study adopts the standard workflow for science mapping concluded by Zupic and Cater [32], which comprises the following five stages: study design, data compilation, data analysis, data visualization, and interpretation. The purpose of this study is to map the research area of household waste recycling and gain insights into this area through bibliometric analysis. Therefore, certain keywords, such as "household waste recycling", were identified as the search terms, and the publication database was filtered using Web of

Science. After finishing the data collection process, the bibliometric method was adopted to carry out a series of data analyses: general descriptive analysis, co-citation analysis, co-word analysis, collaborative network analysis, and thematic evolution analysis. Several scientific publication networks were established from the data and visualized for a better understanding of the research trends. Finally, the results were interpreted and discussed.

2.2. Data Collection

In this study, Clarivate Analytics Web of Science (WoS) was chosen as the source of bibliographic information, as it is one of the most widely utilized databases in scientometrics [33–35] and provides comprehensive and detailed literature data for download [33,36]. Data was obtained from the Science Citation Index Expanded (SCI-E) and the Social Sciences Citation Index (SSCI) databases with the search keywords "Household Waste Recycling". The timespan was limited to all years no later than 2020, and the inquiry was performed on 24 February 2021. Furthermore, the search results were limited to only English articles. As a result, a total of 1295 publications were finally returned, with the very first one published in 1991.

Data exported from the results queried through WoS consists of the following information: author(s), title, source, abstract, cited reference, addresses, times cited, keywords, and cited reference count. These contents were further used as the materials for the bibliometric analysis presented in detail in Section 3.

2.3. Bibliometric Analysis

Bibliometric analysis in this paper mainly entails descriptive analysis and network extraction. Descriptive analysis is used to investigate the evolution of scientific production and identify the most influential articles, authors, journals, and countries of the research area. Network extraction is further divided into different approaches: co-citation analysis, co-author analysis, co-word analysis, and bibliographic coupling analysis. These approaches help us map the social structure of the research community and the conceptual structure of the field.

The R package bibliometrix 3.0.3 developed by Aria and Curulo [24], and the software RStudio were used to perform bibliometric analysis and visualization of the results. At present, researchers have developed numerous tools for bibliometric analysis, the most commonly used ones being VOSviewer [37], SciMAT [38], BibExcel [39], CitNetExplorer [40], and CiteSpace [41]. However, many of these software do not support the complete standard workflow of science mapping [24]. Programmed in the open-source R language, the bibliometrix package cannot only meet all the needs of the whole-process bibliometric analysis but also integrate with other statistical R packages flexibly.

3. Results and Discussion

This section presents the results drawn from the bibliometric analysis and visualizes the related bibliographic networks.

The data collection contains 1295 published articles from 334 journals indexed by SCI-E and SSCI up to 2020, as well as 36,544 references cited by the articles. A total of 3362 authors contribute to the collected publications, of which most are multi-authored articles. The majority of the publications (1048 or 80.93%) fall into the Web of Science categories of environmental sciences (872 or 67.34%), engineering environmental (574 or 44.32%), environmental studies (191 or 14.75%), green sustainable science technology (183 or 14.13%), and economics (89 or 6.87%). In the top 5 WoS categories, 843 articles are multidisciplinary, accounting for 80.44% of the total number.

In the following subsections, we first present a descriptive analysis of scientific production by years, countries, authors, and journals. Some indices are also calculated as a supplement to measure the productivity and citation impact. Then, the social structure of the research area is mapped through collaboration analysis to show how authors, institutions, and countries relate to each other. The third part of this section focuses on the most influential research in the field, including (1) the most cited publications in the dataset; (2) the most cited references in the dataset, which refer to the publications cited the most by the articles in the data collection; (3) the most local cited publications, which refer to the publications in the data collection cited the most by other publications in the collection. The fourth part analyses the high-frequency keywords and major research themes to reveal the conceptual structure of the field. In addition, a Sankey diffluence diagram of different time slices is presented to help identify the evolution of research themes in the field of household waste recycling over time.

3.1. Analysis of Scientific Production

Figure 1 shows the trend in quantity of articles published between 1991 and 2020. The query performed as described in Section 2 returned no result before 1991 as the Web of Science Core Collection mainly started archiving in 1992 [36]. The number of publications followed a mild but continuous growth trend until around 2005, when a more rapid growth began. Starting from 2009, the number of scientific productions fluctuated for four years. A new round of rapid growth started again from 2014 until now. Over the past 30 years, the annual number of publications increased from 1 in 1991 to 141 in 2020, with an annual growth rate of 18.61%. The general trend indicates that this field is still in the expansion period and receives increasing attention from research communities.

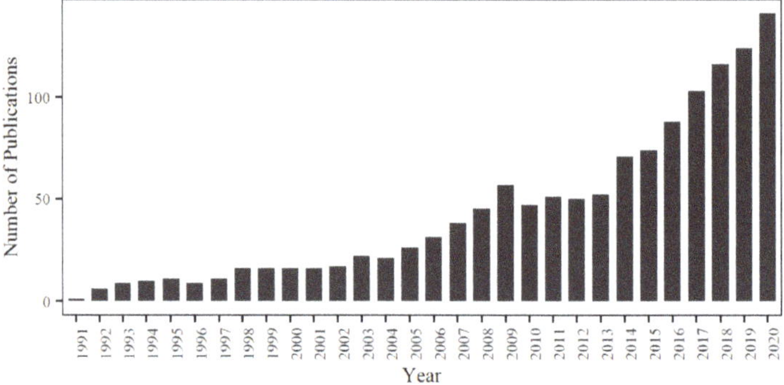

Figure 1. The annual number of publications about household waste recycling based on the data from SCIE and SSCI.

Figure 2 shows the 10 most productive countries and the number of publications by them. As the most populated county and facing serious waste management problems [42–44], China also produced the most publications of all the countries. The United States, who generated more municipal solid waste per year than any other countries [42,45], is the second most productive country regarding the number of publications. The United Kingdom is closely behind the United States. Japan ranks fourth but has the second highest number of multiple country publications.

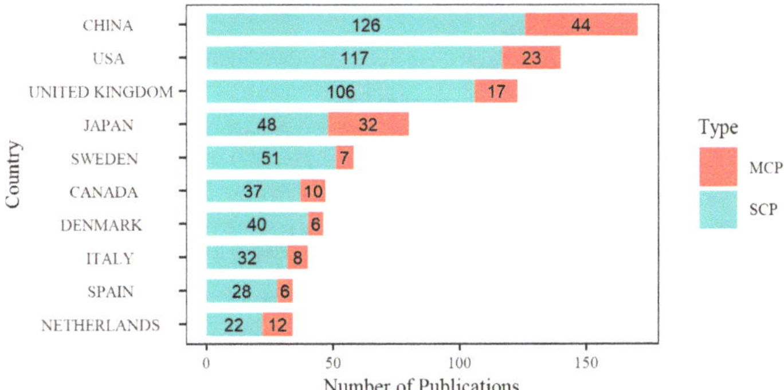

Figure 2. Number of publications in SCIE and SSCI about household waste recycling by country, grouped by multiple country publications (MCP) and single country publications (SCP).

However, when taking into consideration the number of total citations, the United Kingdom and the United States have the highest number of total citations (5343 and 4146, respectively). Germany holds the first position in terms of the average article citations (58.94 citations per article). Although having produced the most publications, China has relatively lower average citations than most of the countries on the list. Table 1 lists the top 10 countries with the highest scientific production in household waste recycling.

Table 1. Top 10 countries ranked by total citations.

Rank	Country	Total Citations	Average Article Citations
1	United Kingdom	5343	43.44
2	USA	4146	29.61
3	China	3353	19.72
4	Sweden	1938	33.41
5	Germany	1827	58.94
6	Italy	1364	34.1
7	Japan	1192	14.9
8	Denmark	1143	24.85
9	Canada	1110	23.62
10	Malaysia	970	33.45

The research field involved 3362 authors in total, among which 2876 authors have 1 article, 440 authors have 2–4 articles, 37 authors have 5–7 articles, and 9 authors have more than 8 articles published.

Table 2 shows the main variables related to the most influential authors on the research topic. Ian Williams from the University of Southampton is the most productive author with 16 published articles (total articles in WoS: 102, overall h-index: 30), followed by Marie Harder with 13 publications (total articles in WoS: 64, overall h-index: 18), and Adam Read with 10 publications (total articles in WoS: 60, overall h-index: 19). The three authors leading the ranking are all affiliated with institutes in the United Kingdom. Four of the other authors on the list are from universities in China, while two are from the Technical University of Denmark. In addition, Ryan Woodard from the University of Brighton co-authored all his eight articles with Marie Harder.

Table 2. Top 10 authors with the highest production of articles.

Authors	Affiliation	h-Index	TC	N	PY
Williams, I.D.	University of Southampton, UK	12	562	16	2004
Harder, M.K.	University of Brighton, UK Fudan University, China	10	240	13	2001
Read, A.D.	University of Northampton, UK	10	790	10	1998
Wang, Y.	Nankai University, China	5	177	9	2010
Astrup, T.F.	Technical University of Denmark, Denmark	9	177	9	2014
Woodard, R.	University of Brighton, UK	8	193	9	2001
Christensen, T.H.	Technical University of Denmark, Denmark	6	236	8	2006
Wang, Z.	Beijing Institute of Technology, China	8	285	8	2012
Wu, Y.	Beijing University of Technology, China	7	173	8	2015
Chen, H.	Jinan University, China	4	74	7	2017

TC: total citations; N: number of publications; PY: year of first publication.

Figure 3 describes the main authors' production over time in the field of household waste recycling. Shan Shan Chung from Hong Kong Baptist University published the first paper about a case study of recycling behavior and the attitude of Hong Kong people in 1994 [46], which is the earliest among the main influential authors. Ian Williams and Maire Harder, as the most productive authors, have started to publish papers in 2004 and 2001, respectively, and are still publishing new papers in recent years. Adam Read, who was active in the field between 1998 and 2008, has the highest number of total citations. Adam Read's paper Using the Theory of Planned Behaviour to Investigate the Determinants of Recycling Behaviour: a Case Study from Brixworth, UK, published in *Resources, Conservation and Recycling* journal, has been cited 364 times [47]. Another paper focusing on the implementation of a marketing communications strategy for curbside recycling, published in the same year, has also got 60 citations [48]. Beginning in the 2010s, scholars from China have become more and more influential in the aspect of scientific productions, including Yan Wang from Nankai University, Zhaohua Wang from Beijing Institute of Technology, Yufeng Wu from Beijing University of Technology, and Hui Chen from Jinan University.

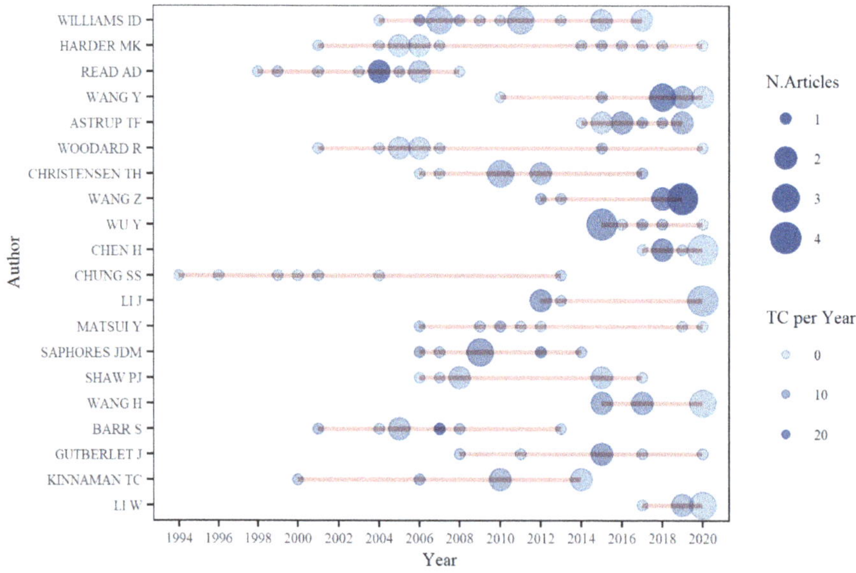

Figure 3. The top 20 authors' production over time. The size of the nodes represents the number of published articles (N.Articles), and the transparency represents the total citations (TC) per year.

A total of 334 journals or other types of sources have published articles on the topic of our study. Of the sources, 69.5% have published only one article, and 16 journals have published 10 or more articles. The top 10 productive journals listed in Table 3 account for 53.2% of the total output. Resources, Conservation and Recycling ranks the first in both the number of articles (180) and total citations (6253), followed by Waste Management (159 articles and 4462 citations), Waste Management & Research (85 articles and 1143 citations), and Journal of Cleaner Production (77 articles and 1831 citations). Resources, Conservation and Recycling also published the first article on household waste recycling in 1991 in our data collection. Besides the journals focusing on waste management, more comprehensive journals about environmental studies also appears on the top list, such as Journal of Cleaner Production, which also publishes articles about cleaner production, environmental, and sustainability research.

Table 3. Top 10 journals with the highest production of articles.

Source	h-Index	TC	N	PY
Resources, Conservation and Recycling	42	6253	180	1991
Waste Management	38	4462	159	2003
Waste Management & Research	18	1143	85	1994
Journal of Cleaner Production	24	1831	77	2002
Sustainability	10	349	53	2012
Journal of Environmental Management	21	1384	39	1996
Journal of Material Cycles and Waste Management	10	208	29	2009
Science of the Total Environment	12	363	27	2008
Environment and Behavior	21	2018	23	1994
Ecological Economics	10	458	17	2002

TC: total citations; N: number of publications; PY: year of first publication.

3.2. Collaboration Network Analysis

Collaboration networks demonstrate how authors, institutions and countries relate to each other, thus helping us to understand the social structure in our field of research. Figure 4 maps the collaboration between the 50 most productive authors. The colors in the figure represent different clusters formed by the academic communities, which indicate a few groups of authors collaborating closely. The size of the nodes refers to the number of co-authored articles published by each author, while the thickness of the links between the nodes indicates the degree of cooperation between certain authors. Five main clusters appear in the network, dominated by Harder MK, Wang H, Williams ID, Chen H, and Astrup TF. The map shows a very strong link between Harder MK and Woodard R, even though they barely collaborate with other authors. Williams ID and Astrup TF also dominate relatively small work groups. As a comparison, the clusters centered on Chen H and Wang H have a higher density and have more authors included. The two groups are also younger than others, as most authors within them started to publish their first articles after 2014. Overall, the collaboration network is discrete as there is no connection between many clusters. Some influential authors like Read AD do not have a stable community to work with, while the younger generations (e.g., Chen H, Wang H) tend to collaborate more with others.

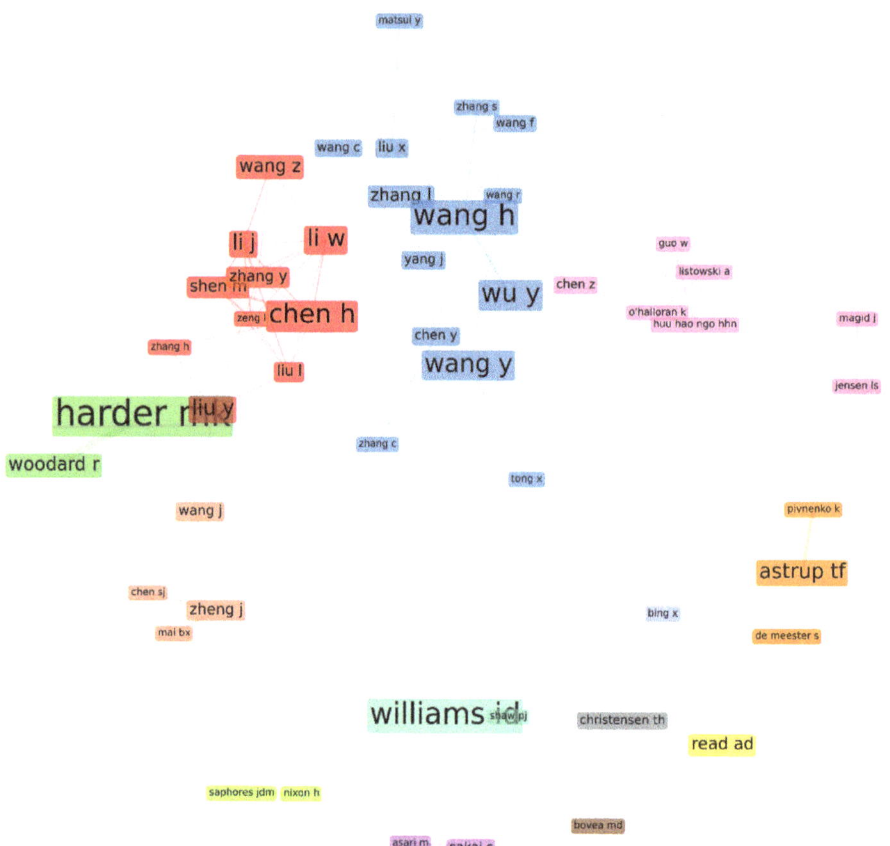

Figure 4. Collaboration network of authors.

Figures 5 and 6 show the collaboration networks of institutions and countries based on co-authorship. The Technical University of Denmark is the most prolific institution in terms of academic cooperation, with 62 publications in total, followed by the University of Southampton (48 articles) and the University of Tokyo (33 articles). The University of Brighton (27 articles), Hong Kong Polytechnic University (25 articles), and Tsinghua University (21 articles) also represent subnetworks of the collaboration map. The University of California, Irvine, ranks third in the number of publications with 36 articles, even though it has collaborated little with other institutions.

The country collaboration map has no obvious cluster comparing to those of authors and institutions, since connecting lines exist between most of the countries. The United Kingdom, China, the United States, and Japan have extraordinary performance in international collaboration. China and the United States have the strongest academic cooperation between each other. The United Kingdom and Japan mainly collaborate with Australia and Vietnam, respectively.

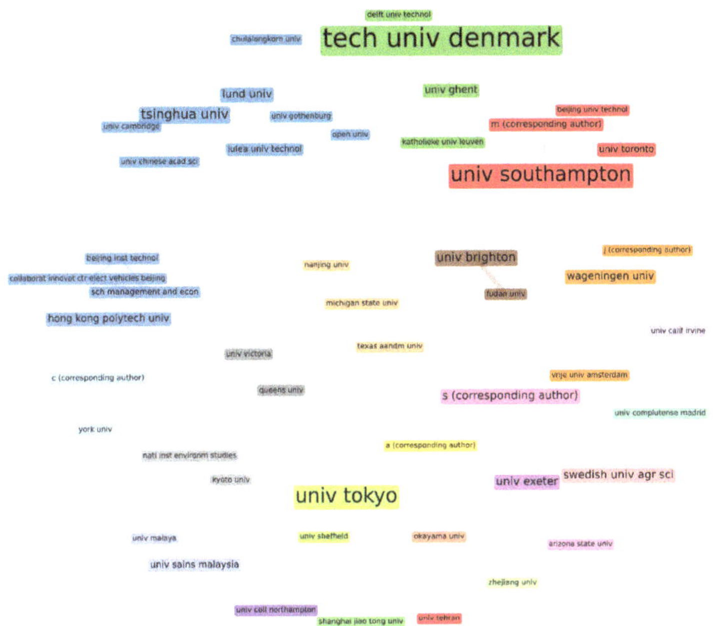

Figure 5. Collaboration network of institutions.

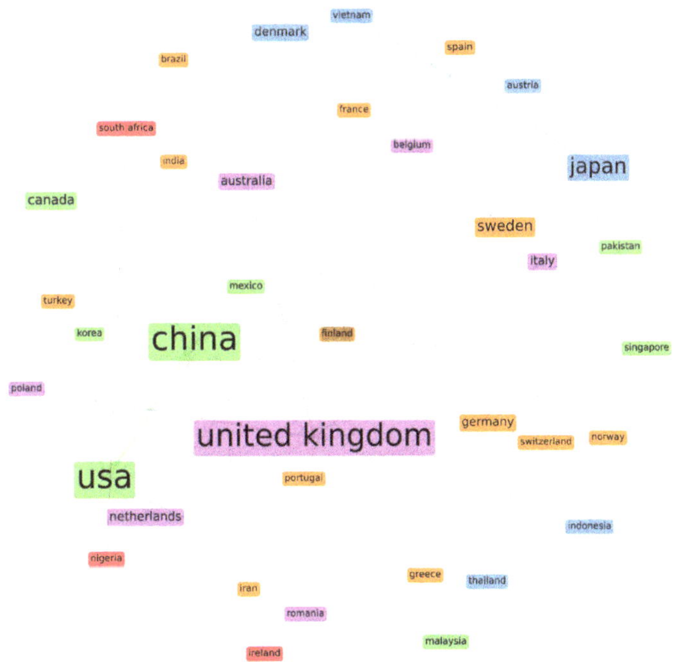

Figure 6. Collaboration network of countries.

3.3. Citation Analysis

Table 4 lists the most cited articles in the dataset. The paper Twenty Years after Hines, Hungerford, and Tomera: A New Meta-analysis of Psycho-social Determinants of Pro-environmental Behaviour, published by Bamberg and Möser [49] in 2007, has drawn the most attention from the academic communities with 1367 total citations. Based on 57 previous empirical studies about correlations between psycho-social variables and pro-environmental behavior, this article conducted a meta-analytic structural equation modelling test and confirmed that personal pro-environmental behavioral intention was determined by attitude, behavior control, personal moral norm, and problem awareness. This intention eventually mediated the impact of all other psycho-social variables on pro-environmental behavior, including household waste recycling.

Table 4. Top 10 global cited articles on household waste recycling.

R	Reference	Title	TC
1	Bamberg and Möser (2007) [49]	Twenty years after Hines, Hungerford, and Tomera: A new meta-analysis of psycho-social determinants of pro-environmental behaviour	1367
2	Tuomela (2000) [50]	Biodegradation of lignin in a compost environment: A review	630
3	Barr (2007) [51]	Factors influencing environmental attitudes and behaviors	367
4	Tonglet et al. (2004) [47]	Using the theory of planned behaviour to investigate the determinants of recycling behaviour: A case study from Brixworth, UK	364
5	Troschinetz and Mihelcic (2009) [3]	Sustainable recycling of municipal solid waste in developing countries	283
6	Quested et al. (2013) [52]	Spaghetti soup: The complex world of food waste behaviours	274
7	Carrus et al. (2008) [53]	Emotions, habits and rational choices in ecological behaviours: The case of recycling and use of public transportation	269
8	Gamba and Oskam (1994) [54]	Factors influencing community residents' participation in commingled curbside recycling programs	259
9	Taylor and Todd (1995) [55]	An integrated model of waste management behavior	220
10	Mannetti et al. (2004) [56]	Recycling: Planned and self-expressive behaviour	211

R: ranking; TC: total citations.

Most of the widely cited articles focus on the influencing factors behind individual's recycling intention or behavior. Environmental values [51], previous recycling experience [57], personal education on waste management [3], anticipated emotions [53], environmental knowledge [54], and personal identity [56] are also identified as significant factors by various studies on household recycling. In addition to the studies that treat household waste as a broad concept, there are also some articles focusing on specific types of waste, such as compostable packages [50] and food waste [52]. The top ten articles ranked by global citations are published in five different journals, within which Journal of Environmental Psychology and Environment and Behavior have both published 3 articles, indicating that psychological methods and behavioral studies are the hotspots in this field.

To further identify the most important articles that have made great contributions within the research field of household waste recycling, local citation is introduced as another indicator to measure the influence. Local citation measures how many times an article included in our data collection has been cited by other articles also in the collection. As listed in Table 5, some of the most local cited articles have a relatively lower number of global citations, indicating that these studies mainly draw attention in the specific research area of household waste recycling. Like the articles with high global citations, many of the most local cited studies also focus on the determinants of recycling intention and behavior. However, some other studies treat recycling from the perspective of economics and policy. Kinnaman and Fullerton [58] estimated the impact of garbage fees and found that correction for endogenous local policy increases the effect of garbage fees on recycling. Dahlen et al. [59] compared the efficiency of different waste collection systems and confirmed that weight-based billing reduced a greater amount of waste than a fixed garbage fee. The rise in the waste collection fee induced residents to recycle more according to the article published by Hong [60] in 1999.

Table 5. Top 10 local citated articles in household waste recycling.

R	Reference	Title	LCS	GCS
1	Gamba and Oskam (1994) [54]	Factors influencing community residents' participation in commingled curbside recycling programs	72	259
2	Bartelings and Sterner (1999) [61]	Household waste management in a Swedish municipality: Determinants of waste disposal, recycling and composting	66	147
3	Knussen et al. (2004) [62]	An analysis of intentions to recycle household waste: The roles of past behaviour, perceived habit, and perceived lack of facilities	59	170
4	Kinnaman and Fullerton (2000) [58]	Garbage and recycling with endogenous local policy	55	108
5	Saphores et al. (2006) [63]	Household willingness to recycle electronic waste	55	144
6	Chan (1998) [64]	Mass communication and pro-environmental behaviour: Waste recycling in Hong Kong	50	166
7	Barr (2007) [51]	Factors influencing environmental attitudes and behaviors	48	367
8	Berglund (2006) [65]	The assessment of households' recycling costs: The role of personal motives	42	84
9	Dahlen et al. (2007) [59]	Comparison of different collection systems for sorted household waste in Sweden	40	88
10	Hong (1999) [60]	The effects of unit pricing system upon household solid waste management: The Korean experience	36	78

R: ranking; LCS: local citations; GCS: global citations.

Since we have already analyzed the most global-cited and local-cited papers, Table 6 lists articles that have been cited the most by the 1295 articles in our data. These articles have either developed theoretic models or provided empirical research examples for the later studies on household waste recycling. The theory of planned behavior (TPB) proposed by Ajzen [66] has been widely adopted to explain individual's recycling intention and behavior. The article about TPB published by Ajzen in 1991 has been cited 108 times by the publications on household waste recycling in our data. TPB assumes that people's intention to perform a certain behavior is mediated by attitude, subjective norm, and perceived behavior control. When applying TPB to household recycling behavior, various researchers introduced additional variables to improve the explanatory power of the theoretic framework, such as moral norm, awareness of consequences [47], past behavior [67], convenience [49], local norm [68], and habits [18].

Table 6. Top 10 articles cited the most by articles in the data.

R	Reference	Title	TC
1	Ajzen (1991) [66]	The theory of planned behavior.	108
2	Vining and Ebreo (1990) [69]	What makes a recycler: A comparison of recyclers and nonrecyclers	94
3	Jenkins et al. (2003) [70]	The determinants of household recycling: A material-specific analysis of recycling program features and unit pricing	85
4	Tonglet et al. (2004) [47]	Using the theory of planned behaviour to investigate the determinants of recycling behaviour: A case study from Brixworth, UK	85
5	Fullerton and Kinnaman (1996) [71]	Household responses to pricing garbage by the bag	80
6	Oskamp et al. (1991) [72]	Factors influencing household recycling behavior	75
7	Gamba and Oskam (1994) [54]	Factors influencing community residents' participation in commingled curbside recycling programs	72
8	Hornik et al. (1995) [73]	Determinants of recycling behavior: A synthesis of research results	70
9	Martin et al. (2006) [23]	Social, cultural and structural influences on household waste recycling: a case study	70
10	Schultz et al. (1995) [74]	Who recycles and when? A review of personal and situational factors	67

R: ranking; TC: total citations.

Finally, we perform a historiographic analysis on the data as proposed by Garfield [75] and plot a chronological direct citation network named historiography [24] (Figure 7). The historiograph helps us quickly identify the most significant works on household waste recycling and trace their year-by-year historical development.

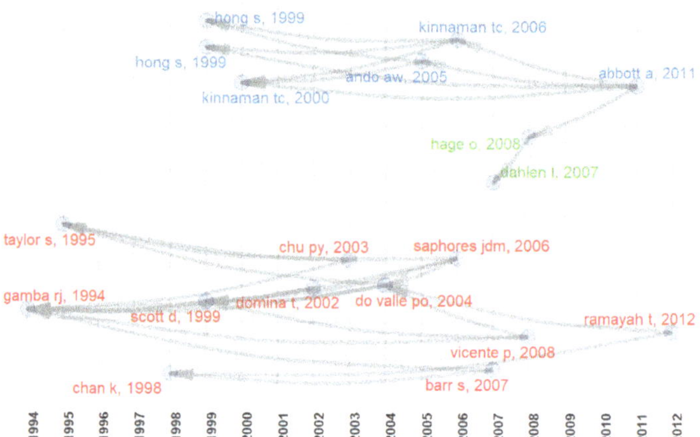

Figure 7. Historiograph top-cited papers in the field of household waste recycling.

The earliest node is the paper Factors Influencing Community Residents' Participation in Commingled Curbside Recycling Programs published by Gamba RJ [54] in 1994. In this paper, a mail survey sent to households in a suburb with a new commingled curbside recycling program discovered an inconsistency between self-reported participation and observed actual recycling behavior. Relevant recycling knowledge was found to be the most significant predictor of observed recycling behavior. Gamba's study has thereafter inspired many of the top-cited articles in the historiograph.

Hong S published two important articles that have triggered four citation chains in 1999. The first paper, The Effects of Unit Pricing System Upon Household Solid Waste Management: The Korean Experience, is a single-author article published in Journal of Environmental Management [60]. The second paper, Household Responses to Price Incentives for Recycling: Some Further Evidence, published in Land Economics, is a collaboration with Adam RM [76]. The two papers both studied the effects of price incentives on household recycling but in different places: 20 cities in South Korea, and Portland, USA. Results from the two empirical studies show that households would recycle more when facing an increase in waste disposal service fees. Meanwhile, according to the study in South Korea, the demand for waste collection services does not necessarily decrease with additional increases in the collection fee unless further recycling incentives are accompanied.

As can be seen in the historiograph, the historical direct citation network of top-cited papers is divided into two subnetworks. One starts from Gamba RJ, 1994, and the other starts from Hong S, 1999. The subnetwork in the lower part of the graph, marked in red, commits to understanding people's recycling intentions and behavior from the perspective of social psychology, while articles in the upper part marked in blue apply economic models to studies of garbage fees. This dispersion observed from the historiograph indicates the two main perspectives on the research of household waste recycling. Some researchers also interacted with the two perspectives in their studies. As marked in green in Figure 7, the paper Norms and Economic Motivation in Household Recycling: Empirical Evidence from Sweden, published by Hage O [77] in 2008, built a theoretical framework that integrated norm-motivated behavior into an economic model and confirmed that both economic and moral motives could influence household recycling rates.

3.4. Conceptual Structure Analysis

3.4.1. Descriptive Analysis of High-Frequency Keywords

Keywords are the high-level summarization and refinement of the article core [78]. There are 3545 keywords provided by the authors, and 67 keywords meet the threshold of 10 occurrences. Of the keywords, 2835 only appear once, accounting for 80% of the total amounts, which reflects great diversity of the research field. The most frequent keyword, "recycling", occurs in 343 articles, as it is included in the searching keywords in the data collection process. To eliminate the interference of repeated keywords and more accurately identify research hotspots, we exclude the homogeneous keywords that are highly similar to our search keywords, as well as the overly broad concepts concerning environmental studies. The excluded keywords and their occurrences are as follows: "recycling" (343), "waste management" (108), "waste" (88), "household waste" (69), "municipal solid waste" (50), "solid waste" (36), "solid waste management" (30), "management" (27), "waste recycling" (25), "household" (22), "analysis" (19), "environment" (18), "environmental" (17), "waste collection" (17), "household recycling" (16), "household solid waste" (15), "households" (15), "survey" (14), "municipal waste" (12), "municipal solid waste management" (10), and "recycle" (10). Some other keywords with similar meanings are merged: "sustainability" and "sustainable development"; "food waste" and "organic waste"; "recycling behavior", "recycling behaviour", and "behaviour"; "life cycle assessment", "LCA", and "material flow analysis"; "WEEE", "e-waste", and "electronic waste"; "source separation", "waste separation", "separation", and "waste sorting"; "plastic waste", "plastics", and "plastic recycling"; "policy" and "environmental policy".

Table 7 lists the high-frequency keywords after filtering. As can be seen from the list, the research field of household waste recycling has been extended to all other processes in the waste management chain, including waste generation (11 occurrences), waste collection (31 occurrences), source separation (57 occurrences), and disposal (11 occurrences). In this context, life cycle assessment, defined as "a technique to compile and analyze the environmental impacts involved in all stages of the product's life cycle from raw material extraction stage to the disposal stage" [79], is widely used in waste recycling studies.

Table 7. Most relevant keywords meeting the threshold of 10 occurrences.

Keyword	Occurrences	Keyword	Occurrences
electronic waste	69	theory of planned behavior	17
source separation	57	landfill	14
life cycle assessment	56	reverse logistics	14
recycling behavior	52	social norms	13
sustainability	45	willingness to pay	13
organic waste	39	disposal	11
circular economy	37	questionnaire survey	11
plastic waste	33	recovery	11
attitudes	29	Vietnam	11
composting	26	waste generation	11
China	25	consumption	10
policy	22	informal sector	10
incineration	21	participation	10
packaging waste	20	urban	10
reuse	19	waste reduction	10

The following terms in the table are merged with the keywords in the brackets: source separation (waste separation, separation, waste sorting); life cycle assessment (LCA, material flow analysis); recycling behavior (recycling behaviour, behaviour); sustainability (sustainable development); organic waste (food waste); electronic waste (WEEE, e-waste); plastic waste (plastics, plastic recycling); attitudes (attitude); policy (environmental policy).

Sustainable development and the circular economy are two important concepts that drive the studies of waste recycling. Household waste recycling as a part of municipal solid waste management is the foundation for a circular economy to achieve more waste prevention and better resource management. Even though various studies have discussed waste recycling in the context of the circular economy and sustainable development, the relationships between them are still blurred as these concepts have been diffuse since they were proposed [30]

The keyword list also shows that personal recycling behavior is the most popular research object of the articles. The theory of planned behavior is widely used in relevant behavioral studies. Attitude and social norm are the most studied predictors of recycling behavior.

In all types of household waste, waste from electrical and electronic equipment (WEEE), food waste and organic waste, plastic waste, and packaging waste are mentioned the most by the authors. Waste reduction and reuse, as another two elements of the three Rs (reduce, reuse, recycle), have been discussed frequently with household recycling.

Despite the original keywords provided by the authors, our data also includes another type of keywords, namely keywords plus. Keywords plus are generated from the titles of an article's references based upon a unique algorithm [80]. Compared with authors' keywords, keywords plus are more broadly descriptive but less comprehensive in presenting the content of a specific article [81]. Figure 8 presents the yearly occurrences of the top 10 keywords plus terms. Household waste recycling has been treated mostly as a management issue, as the term "management" occurs a lot more than the other terms during the past 10 years. "Recycling behavior" draws little attention around 2010, but since then, the number of its occurrences has increased the most rapidly. On the contrary, "participation" is the only one of the 10 most relevant keywords plus terms that has declined in recent years.

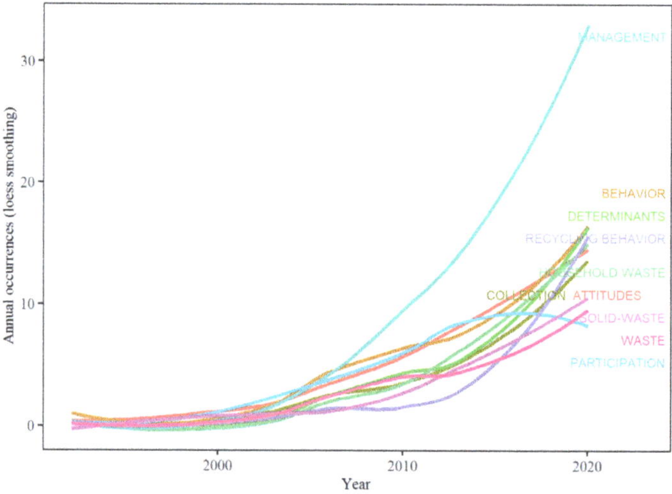

Figure 8. Annual occurrences of the most relevant keywords plus terms generated by WoS.

3.4.2. Cluster Analysis of High-Frequency Keywords

Using multiple correspondence analysis, the conceptual structure maps of authors' keywords and keywords plus are generated, and the keywords clusters are plotted in two-dimensional maps (Figures 9 and 10). The closer the points representing each keyword are on the graph, the more similar the distribution of the keywords are, which means they co-occur in the articles more frequently. Moreover, the proximity of a keyword to the center point represents its popularity in the research field. Keywords around the center have received high attention from the research community, while those by the edge are less related to other research topics [82].

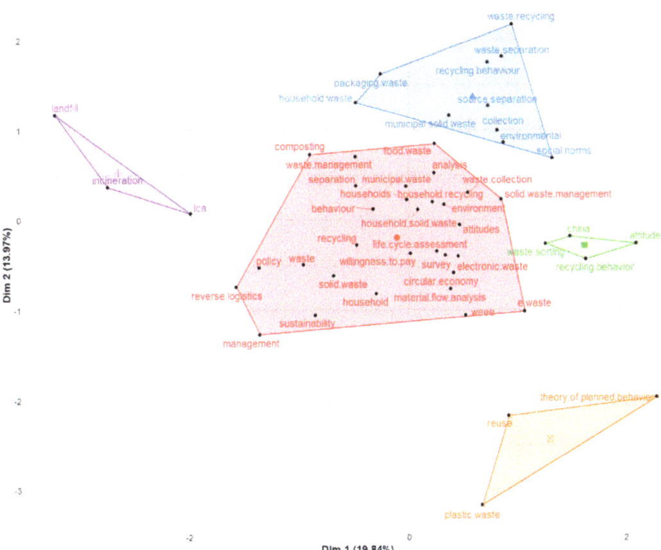

Figure 9. Cluster analysis of the authors' keywords using multiple correspondence analysis.

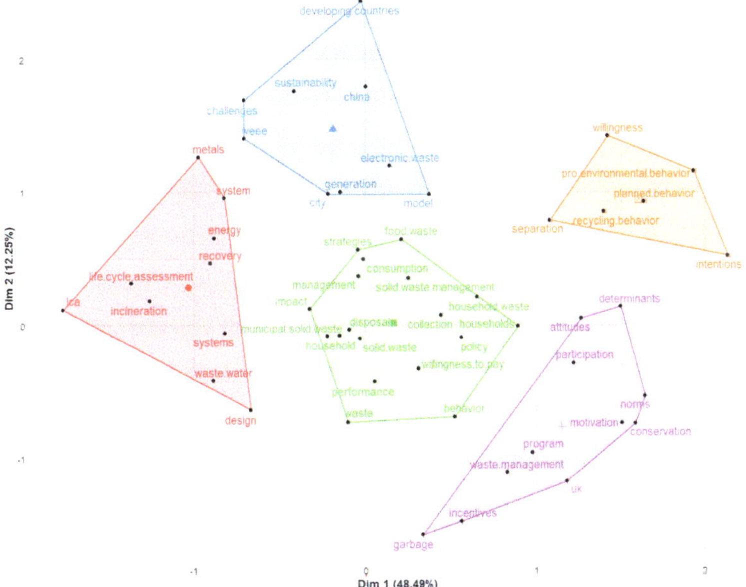

Figure 10. Cluster analysis of the keywords plus using multiple correspondence analysis.

In terms of authors' keywords, we can infer from Figure 9 that the keywords are grouped into two major clusters and three minor clusters:

(1) The first major cluster marked in red in the center of the graph involves most of the important concepts regarding the research field and is highly consistent with the topic of household waste recycling. "Behavior", "attitudes", and "willingness to pay" are the

main research variables of this group of studies. "Life cycle assessment", "material flow analysis", and "reverse logistics" have also been emphasized in this cluster. Apart from the general sense of household waste, more specific types of waste, including food waste and electronic waste, have attracted considerable research interest.

(2) The second major cluster, which is marked in blue and located in the upper right part of the graph, is more related to studies on waste source separation.

(3) The first minor cluster, marked in purple on the left side of the graph, is dedicated to the study of traditional waste disposal methods, including landfill and incineration. Life cycle assessment (LCA) is a commonly used technique in these kinds of studies.

(4) The second cluster, marked in green on the right side of the graph, is mainly related to waste sorting, recycling behavior, attitude, and research carried out in China.

(5) The last minor cluster, marked in yellow on the bottom right, is mainly related to the reuse of plastic waste and the application of the theory of planned behavior.

Overall, Figure 9 shows the distribution pattern of authors' keywords, demonstrating that household waste recycling, as the core research topic, is surrounded by other related topics such as source separation, reuse, and final disposal.

The co-occurrence of the keywords plus is more uniformly distributed. As can be seen from Figure 10, five clusters with similar scales are generated using multiple correspondence analysis:

(1) The first cluster, marked in red, focuses more on the technical and engineering aspects. High-frequency keywords that dominate this cluster are life cycle assessment (LCA), incineration, recovery, systems, and energy.

(2) The second cluster, marked in blue, focuses on electronic waste (WEEE) and sustainability. The research areas of the studies in this cluster are mainly China and other developing countries.

(3) The third cluster, marked in green, is dedicated to household solid waste management. Collection, consumption, willingness to pay, performance, and disposal are important topics in this group of studies.

(4) The fourth cluster, marked in yellow, is related to numerous studies that apply the theory of planned behavior. This popular theory in the field of pro-environmental behavior is not only applicable to recycling behavior but also widely used in studies of source separation.

(5) The last cluster, marked in purple, is mainly related to concepts and topics under social psychology. Top contributing keywords in this cluster are motivation, norms, incentives, attitude, program, and determinants. The United Kingdom (UK) also appears in this cluster, indicating that many of the studies from the perspective of social psychology are contributed by British scholars.

3.4.3. Thematic Evolution Analysis

The clusters of keywords obtained from the co-word analysis are considered as themes of our research field [83]. We further created a thematic map (Figure 11) based on co-word network analysis and clustering. The thematic map, also referred to as the strategic diagram [84,85], describes two parameters ("centrality" and "density") that characterize the themes in a two-dimensional space. Centrality measures to what extent a network interacts with other networks. The centrality of a theme represents the strength of its external connections to other themes and can be used as an indicator to measure the influence of the theme in the entire research field. Density measures the strength of internal ties among all the keywords within a theme. Therefore, the density of a given theme represents its development.

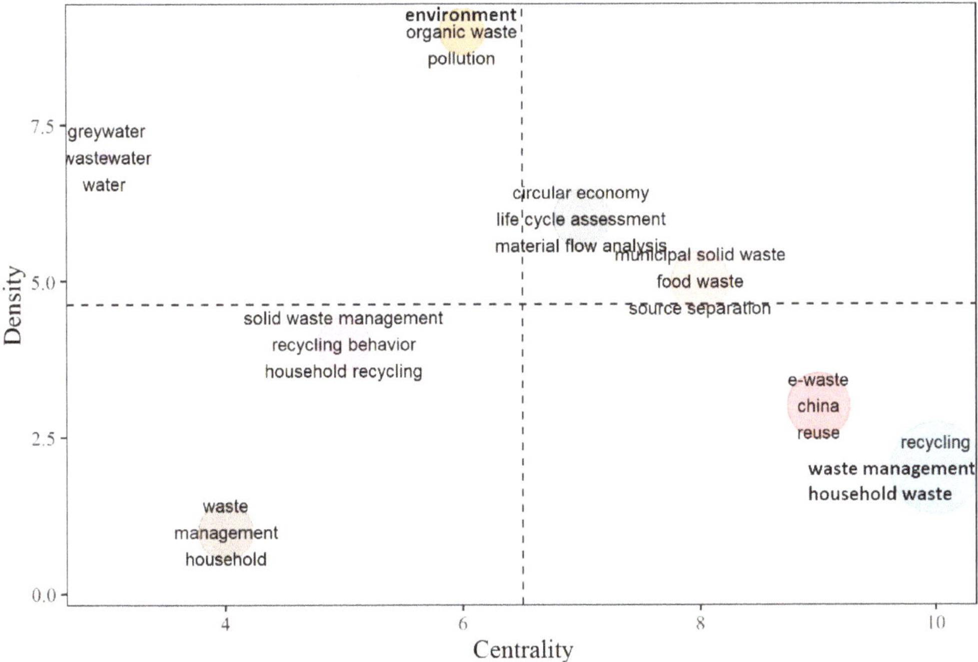

Figure 11. Thematic map of authors' keywords based on co-word network analysis and clustering.

Based on their centrality and density, the themes distributed in four quadrants in Figure 11 are defined by the following four categories [83,84,86,87]:

- Motor themes in the upper-right quadrant. Such themes are both fully developed and vital to the research field.
- Specialized and peripheral themes in the upper left quadrant. Given that these themes have a relatively higher density but lower centrality, they are isolated and have limited influence on the field despite their distinctive internal development.
- Emerging or declining themes in the lower left quadrant. The themes of this category are weakly developed and marginal to the research field.
- Basic and transversal themes in the lower right quadrant. They are not yet fully developed, but they have a very important position in the field of research.

According to the strategic diagram generated using authors' keywords from our data, we can observe eight main themes with different levels of density and centrality. What stands out is that different types of waste differ greatly in their positions in the figure. Electronic waste (e-waste), belonging to the basic and transversal themes, has a much higher centrality but lower density compared with other types of waste. Greywater has the lowest centrality, implying that this specialized theme is relatively peripheral and marginal in our field of research. The theme represented by "circular economy", "life cycle assessment", and "material flow analysis" has its centrality and density both above the average line. However, the upper right part of the strategic diagram is still vacant, indicating that motor themes still need to be found in future studies.

In the last part of this section, we perform thematic evolution analysis and map the results with a Sankey diagram (Figure 12). The Sankey diagram helps us clarify the quantity and direction of thematic flow and conversion relationships between the themes [88].

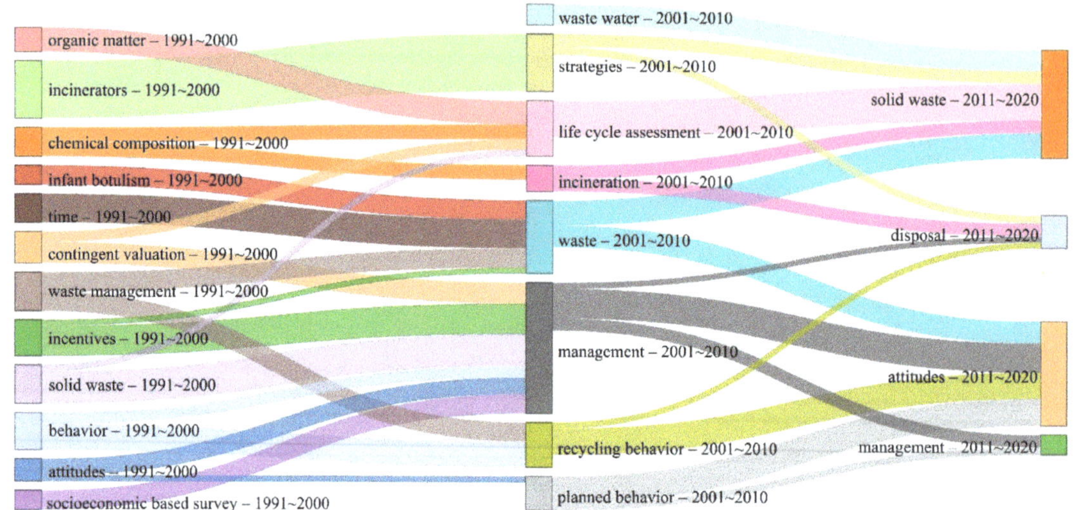

Figure 12. Thematic evolution of the research field of household waste recycling (1991–2020).

The timespan of our data is evenly divided into three slices with two cutting years, 2000 and 2010. The nodes in the diagram represent the main research themes generated form the co-word network analysis in each time slice. The text labels next to the nodes indicate the core keywords of the themes as well as the time slices. The number of keywords included in each theme is represented by the size of the corresponding node. Themes from adjacent time slices are connected by streamlines when they share the same keywords. The width of the streamlines is proportional to the number of keywords shared by the connected themes and indicates the relevance between them.

From Figure 12, we can see that as the research progresses, the pattern of research themes has gradually shifted from decentralization to uniformity. In the early stage (1991–2000), various themes are dominated by specialized and unrelated keywords, such as "incinerators", "chemical composition", "contingent valuation", "incentives", and "behavior". In the second stage (2001–2010), the methodology of life cycle assessment and the theory of planned behavior have been established, developed, and widely used in the research field. Under the theme of "management", numerous interdisciplinary studies have emerged from isolated themes, including "contingent valuation", "incentives", "solid waste", "behavior", "attitudes", and "socioeconomic based survey" in the first stage. In the third stage (2011–2020), the comprehensiveness of the research field is further enhanced, as different themes from the former time slices merge again into new themes.

4. Conclusions

Municipal solid waste problems have attracted increasing concerns, and several studies have systematically reviewed the field of related topics. Since household waste recycling is a key concept in municipal waste management and there is still lack of scientometric studies focusing on it, this paper performed a bibliometric analysis based on the data of 1295 scientific publications obtained from Web of Science using the open-source R language and bibliometrix package.

The analysis of scientific productivity shows that the research field of household waste recycling is still in the expansion period as annual rapid growth has not slowed down. China is the most productive country in terms of the total number of articles published. However, when taking into consideration the number of citations as an indicator of academic influence, the United Kingdom and the USA have surpassed China. Ian Williams, Marie Harder, and Adam Read are the most prolific authors, and they are all affiliated with institutes in the UK. Beginning in the 2010s, researchers from China has become increasingly influential in terms of both the quantity and the quality of scientific productions, challenging the dominance of British scholars. Resources, Conservation and Recycling and Waste Management published influential studies on household waste recycling.

Collaborative network analysis reveals that younger generations tend to cooperate more with each other in their studies than senior scholars. As the two countries with the most academic output, China and the USA also have the closest cooperation relationship, while the UK mainly collaborates with European countries. Among all the institutions, the Technical University of Denmark is the most prolific in terms of academic cooperation.

Citation analysis helps us identify the most influential articles in our research field. A meta-analysis of determinants of pro-environmental behavior, including household waste recycling carried out by Bamberg and Möser [49], has received the greatest interest and got the most citations. In terms of studies not included in our data, the theory of planned behavior, proposed by Ajzen [66], has made the greatest contribution to our field of research. Numerous studies applied this theory to the identification of factors influencing residents' recycling behavior. In general, most of the highly cited articles either focus on socio-psychological or economic perspectives. This trend is also confirmed by historiographic analysis.

The conceptual structure analysis first identifies research hotspots including electronic waste, source separation, life cycle assessment, recycling behavior, and sustainability. Then, keywords are clustered into themes using multiple correspondence analysis, and a strategic diagram is generated. According to the strategic diagram, the circular economy and life cycle assessment are clarified as motor themes that are fully developed and vital to the research field. Organic waste and pollution are specialized and peripheral themes. Electronic waste and reuse are basic and transversal themes that still need further development. Finally, a Sankey diagram generated from thematic evolution analysis demonstrates the patterns of the three stages of thematic evolution. Research themes are specialized and isolated in the first stage (1991–2000). Methodologies and theories are developed and widely applied in the second stage (2001–2010), including life cycle assessment and the theory of planned behavior. In the third stage (2011–2020), interdisciplinary trends are observed as several dispersed themes from previous stages merge into new comprehensive themes.

Our study also contributes to the literature given that new findings have been made with comparisons to previous bibliometric studies in the field of waste management. Li et al. [25] identified pyrolysis of e-waste, biodiesel production from waste oil, and anaerobic digestion of organic waste as the research hotspots in their study of "solid waste reuse and recycling". Tsai et al. [30] concluded that incineration is the top indicator for future study of "municipal solid waste in a circular economy". When focusing on household waste recycling, our study finds out that besides these technological topics, research on personal recycling behavior from the perspective of psychology has attracted even more attention. Household waste recycling is not only an environmental science issue but also a social science issue.

With the findings discussed above, this article may help clarify the current research status and future directions of household waste recycling and have implications for both public authorities and academics. Governing bodies should assess the environmental impacts of different types of waste disposal from the entire life cycle and establish sustainable household waste recycling systems under the guidance of academic research. The influencing factors behind residents' attitudes and willingness to recycle must be seriously considered in the policy-making process. Researchers should pay attention to and seek opportunities for collaboration with institutes and countries that are experts in their specific area of interest, such as China for e-waste studies and the United Kingdom for incentives and personal behavior. Communication and collaboration between institutes need to be strengthened as very limited cooperation has been observed. Finally, despite the interdisciplinary trends appearing in the thematic evolution process, research perspectives are still limited as many newly published studies are just repetitions of old ones. Frontier and innovative research need to be explored to push the boundaries of waste management and sustainable development.

Author Contributions: Conceptualization, K.S.; methodology, K.S.; software, K.S.; validation, K.S. and Y.Z.; data curation, K.S.; writing—original draft preparation, K.S.; writing—review and editing, K.S., Y.Z., and Z.Z.; visualization, K.S.; supervision, Z.Z.; project administration, Z.Z. All authors have read and agreed to the published version of the manuscript.

Funding: This research received no external funding.

Institutional Review Board Statement: Not applicable.

Informed Consent Statement: Not applicable.

Data Availability Statement: Web of Science (WoS).

Conflicts of Interest: The authors declare no conflict of interest.

References

1. Ma, J.; Hipel, K.W. Exploring social dimensions of municipal solid waste management around the globe—A systematic literature review. *Waste Manag.* **2016**, *56*, 3–12. [CrossRef] [PubMed]
2. Wang, H.; Liu, X.; Wang, N.; Zhang, K.; Wang, F.; Zhang, S.; Wang, R.; Zheng, P.; Matsushita, M. Key factors influencing public awareness of household solid waste recycling in urban areas of China: A case study. *Resour. Conserv. Recycl.* **2020**, *158*, 104813. [CrossRef]
3. Troschinetz, A.M.; Mihelcic, J.R. Sustainable recycling of municipal solid waste in developing countries. *Waste Manag.* **2009**, *29*, 915–923. [CrossRef] [PubMed]
4. Ma, J.; Hipel, K.W.; Hanson, M.L.; Cai, X.; Liu, Y. An analysis of influencing factors on municipal solid waste source-separated collection behavior in Guilin, China by using the theory of planned behavior. *Sustain. Cities Soc.* **2018**, *37*, 336–343. [CrossRef]
5. Noor, T.; Javid, A.; Hussain, A.; Bukhari, S.M.; Ali, W.; Akmal, M.; Hussain, S.M. *Types, Sources and Management of Urban Wastes*; Elsevier: Amsterdam, The Netherlands, 2020; pp. 239–263. [CrossRef]
6. Karak, T.; Bhagat, R.M.; Bhattacharyya, P. Municipal solid waste generation, composition, and management: The world scenario. *Crit. Rev. Environ. Sci. Technol.* **2012**, *42*, 1509–1630. [CrossRef]
7. Wang, H.; Nie, Y. Municipal solid waste characteristics and management in China. *J. Air Waste Manag. Assoc.* **2001**, *51*, 250–263. [CrossRef]
8. Knickmeyer, D. Social factors influencing household waste separation: A literature review on good practices to improve the recycling performance of urban areas. *J. Clean. Prod.* **2020**, *245*, 118605. [CrossRef]
9. Gunaratne, T.; Krook, J.; Andersson, H. Current practice of managing the waste of the waste: Policy, market, and organisational factors influencing shredder fines management in Sweden. *Sustainability* **2020**, *12*, 9540. [CrossRef]
10. Saphores, J.-D.M.; Nixon, H. How effective are current household recycling policies? Results from a national survey of U.S. households. *Resour. Conserv. Recycl.* **2014**, *92*, 1–10. [CrossRef]
11. Zheng, P.; Zhang, K.; Zhang, S.; Wang, R.; Wang, H. The door-to-door recycling scheme of household solid wastes in urban areas: A case study from Nagoya, Japan. *J. Clean. Prod.* **2017**, *163*, S366–S373. [CrossRef]
12. Nelles, M.; Grünes, J.; Morscheck, G. Waste management in Germany—Development to a sustainable circular economy? *Procedia Environ. Sci.* **2016**, *35*, 6–14. [CrossRef]
13. Wang, H.; Jiang, C. Local nuances of authoritarian environmentalism: A legislative study on household solid waste sorting in China. *Sustainability* **2020**, *12*, 2522. [CrossRef]
14. Wan, C.; Shen, G.Q.; Yu, A. Key determinants of willingness to support policy measures on recycling: A case study in Hong Kong. *Environ. Sci. Policy* **2015**, *54*, 409–418. [CrossRef]
15. Lizin, S.; Van Dael, M.; Van Passel, S. Battery pack recycling: Behaviour change interventions derived from an integrative theory of planned behaviour study. *Resour. Conserv. Recycl.* **2017**, *122*, 66–82. [CrossRef]
16. Botetzagias, I.; Dima, A.-F.; Malesios, C. Extending the theory of planned behavior in the context of recycling: The role of moral norms and of demographic predictors. *Resour. Conserv. Recycl.* **2015**, *95*, 58–67. [CrossRef]
17. Miafodzyeva, S.; Brandt, N. Recycling behaviour among householders: Synthesizing determinants via a meta-analysis. *Waste Biomass Valorization* **2012**, *4*, 221–235. [CrossRef]
18. Aboelmaged, M. E-waste recycling behaviour: An integration of recycling habits into the theory of planned behaviour. *J. Clean. Prod.* **2021**, *278*, 124182. [CrossRef]
19. White, K.M.; Hyde, M.K. The role of self-perceptions in the prediction of household recycling behavior in Australia. *Environ. Behav.* **2011**, *44*, 785–799. [CrossRef]
20. Khan, F.; Ahmed, W.; Najmi, A. Understanding consumers' behavior intentions towards dealing with the plastic waste: Perspective of a developing country. *Resour. Conserv. Recycl.* **2019**, *142*, 49–58. [CrossRef]

21. Wang, Z.; Guo, D.; Wang, X. Determinants of residents' e-waste recycling behaviour intentions: Evidence from China. *J. Clean. Prod.* **2016**, *137*, 850–860. [CrossRef]
22. Hicks, C.; Dietmar, R.; Eugster, M. The recycling and disposal of electrical and electronic waste in China—Legislative and market responses. *Environ. Impact Assess. Rev.* **2005**, *25*, 459–471. [CrossRef]
23. Martin, M.; Williams, I.D.; Clark, M. Social, cultural and structural influences on household waste recycling: A case study. *Resour. Conserv. Recycl.* **2006**, *48*, 357–395. [CrossRef]
24. Aria, M.; Cuccurullo, C. Bibliometrix: An R-tool for comprehensive science mapping analysis. *J. Informetr.* **2017**, *11*, 959–975. [CrossRef]
25. Li, N.; Han, R.; Lu, X. Bibliometric analysis of research trends on solid waste reuse and recycling during 1992–2016. *Resour. Conserv. Recycl.* **2018**, *130*, 109–117. [CrossRef]
26. Wu, H.; Zuo, J.; Zillante, G.; Wang, J.; Yuan, H. Construction and demolition waste research: A bibliometric analysis. *Archit. Sci. Rev.* **2019**, *62*, 354–365. [CrossRef]
27. Liu, Y.; Sun, T.; Yang, L. Evaluating the performance and intellectual structure of construction and demolition waste research during 2000–2016. *Environ. Sci. Pollut. Res. Int.* **2017**, *24*, 19259–19266. [CrossRef] [PubMed]
28. Gao, Y.; Ge, L.; Shi, S.; Sun, Y.; Liu, M.; Wang, B.; Shang, Y.; Wu, J.; Tian, J. Global trends and future prospects of e-waste research: A bibliometric analysis. *Environ. Sci. Pollut. Res. Int.* **2019**, *26*, 17809–17820. [CrossRef] [PubMed]
29. Zhang, L.; Geng, Y.; Zhong, Y.; Dong, H.; Liu, Z. A bibliometric analysis on waste electrical and electronic equipment research. *Environ. Sci. Pollut. Res. Int.* **2019**, *26*, 21098–21108. [CrossRef]
30. Tsai, F.M.; Bui, T.-D.; Tseng, M.-L.; Lim, M.K.; Hu, J. Municipal solid waste management in a circular economy: A data-driven bibliometric analysis. *J. Clean. Prod.* **2020**, *275*, 124132. [CrossRef]
31. Wang, C.; Liu, D.; Li, Y.; Wang, L.; Gu, W. A multidisciplinary perspective on the evolution of municipal waste management through text-mining: A mini-review. *Waste Manag. Res.* **2020**, 734242X20962841. [CrossRef]
32. Zupic, I.; Čater, T. Bibliometric methods in management and organization. *Organ. Res. Methods* **2014**, *18*, 429–472. [CrossRef]
33. Medina-Mijangos, R.; Seguí-Amórtegui, L. Research trends in the economic analysis of municipal solid waste management systems: A bibliometric analysis from 1980 to 2019. *Sustainability* **2020**, *12*, 8509. [CrossRef]
34. He, H.; Dyck, M.; Lv, J. The heat pulse method for soil physical measurements: A bibliometric analysis. *Appl. Sci.* **2020**, *10*, 6171. [CrossRef]
35. Cañas-Guerrero, I.; Mazarrón, F.R.; Pou-Merina, A.; Calleja-Perucho, C.; Díaz-Rubio, G. Bibliometric analysis of research activity in the "Agronomy" category from the web of science, 1997–2011. *Eur. J. Agron.* **2013**, *50*, 19–28. [CrossRef]
36. Lv, W.; Zhao, X.; Wu, P.; Lv, J.; He, H. A scientometric analysis of worldwide intercropping research based on web of science database between 1992 and 2020. *Sustainability* **2021**, *13*, 2430. [CrossRef]
37. van Eck, N.J.; Waltman, L. Software survey: VOSviewer, a computer program for bibliometric mapping. *Scientometrics* **2010**, *84*, 523–538. [CrossRef]
38. Cobo, M.J.; López-Herrera, A.G.; Herrera-Viedma, E.; Herrera, F. SciMAT: A new science mapping analysis software tool. *J. Am. Soc. Inf. Sci. Technol.* **2012**, *63*, 1609–1630. [CrossRef]
39. Persson, O.; Danell, R.; Schneider, J.W. How to use Bibexcel for various types of bibliometric analysis. *Celebr. Sch. Commun. Stud.* **2009**, *5*, 9–24.
40. van Eck, N.J.; Waltman, L. CitNetExplorer: A new software tool for analyzing and visualizing citation networks. *J. Informetr.* **2014**, *8*, 802–823. [CrossRef]
41. Chen, C. CiteSpace II: Detecting and visualizing emerging trends and transient patterns in scientific literature. *J. Am. Soc. Inf. Sci. Technol.* **2006**, *57*, 359–377. [CrossRef]
42. Zhang, D.Q.; Tan, S.K.; Gersberg, R.M. Municipal solid waste management in China: Status, problems and challenges. *J. Environ. Manag.* **2010**, *91*, 1623–1633. [CrossRef] [PubMed]
43. Wang, Y.; Long, X.; Li, L.; Wang, Q.; Ding, X.; Cai, S. Extending theory of planned behavior in household waste sorting in China: The moderating effect of knowledge, personal involvement, and moral responsibility. *Environ. Dev. Sustain.* **2020**. [CrossRef]
44. Zhuang, Y.; Wu, S.W.; Wang, Y.L.; Wu, W.X.; Chen, Y.X. Source separation of household waste: A case study in China. *Waste Manag.* **2008**, *28*, 2022–2030. [CrossRef] [PubMed]
45. Silpa, K.; Lisa, C.Y.; Perinaz, B.-T.; Frank Van, W. *What a Waste 2.0*; The World Bank: Washington, WA, USA, 2018.
46. Chung, S.S.; Poon, C.S. Recycling behaviour and attitude: The case of the Hong Kong people and commercial and household wastes. *Int. J. Sustain. Dev. World Ecol.* **2009**, *1*, 130–145. [CrossRef]
47. Tonglet, M.; Phillips, P.S.; Read, A.D. Using the Theory of Planned Behaviour to investigate the determinants of recycling behaviour: A case study from Brixworth, UK. *Resour. Conserv. Recycl.* **2004**, *41*, 191–214. [CrossRef]
48. Mee, N.; Clewes, D.; Phillips, P.S.; Read, A.D. Effective implementation of a marketing communications strategy for kerbside recycling: A case study from Rushcliffe, UK. *Resour. Conserv. Recycl.* **2004**, *42*, 1–26. [CrossRef]
49. Bamberg, S.; Möser, G. Twenty years after Hines, Hungerford, and Tomera: A new meta-analysis of psycho-social determinants of pro-environmental behaviour. *J. Environ. Psychol.* **2007**, *27*, 14–25. [CrossRef]
50. Barr, S. Factors influencing environmental attitudes and behaviors. *Environ. Behav.* **2007**, *39*, 435–473. [CrossRef]
51. Tonglet, M.; Phillips, P.S.; Bates, M.P. Determining the drivers for householder pro-environmental behaviour: Waste minimisation compared to recycling. *Resour. Conserv. Recycl.* **2004**, *42*, 27–48. [CrossRef]

52. Carrus, G.; Passafaro, P.; Bonnes, M. Emotions, habits and rational choices in ecological behaviours: The case of recycling and use of public transportation. *J. Environ. Psychol.* **2008**, *28*, 51–62. [CrossRef]
53. Gamba, R.J.; Oskamp, S. Factors influencing community residents' participation in commingled curbside recycling programs. *Environ. Behav.* **2016**, *26*, 587–612. [CrossRef]
54. Mannetti, L.; Pierro, A.; Livi, S. Recycling: Planned and self-expressive behaviour. *J. Environ. Psychol.* **2004**, *24*, 227–236. [CrossRef]
55. Tuomela, M. Biodegradation of lignin in a compost environment: A review. *Bioresour. Technol.* **2000**, *72*, 169–183. [CrossRef]
56. Quested, T.E.; Marsh, E.; Stunell, D.; Parry, A.D. Spaghetti soup: The complex world of food waste behaviours. *Resour. Conserv. Recycl.* **2013**, *79*, 43–51. [CrossRef]
57. Taylor, S.; Todd, P. An integrated model of waste management behavior. *Environ. Behav.* **2016**, *27*, 603–630. [CrossRef]
58. Kinnaman, T.C.; Fullerton, D. Garbage and recycling with endogenous local policy. *J. Urban. Econ.* **2000**, *48*, 419–442. [CrossRef]
59. Dahlen, L.; Vukicevic, S.; Meijer, J.E.; Lagerkvist, A. Comparison of different collection systems for sorted household waste in Sweden. *Waste Manag.* **2007**, *27*, 1298–1305. [CrossRef]
60. Hong, S. The effects of unit pricing system upon household solid waste management: The Korean experience. *J. Environ. Manag.* **1999**, *57*, 1–10. [CrossRef]
61. Bartelings, H.; Sterner, T. Household waste management in a swedish municipality: Determinants of waste disposal, recycling and composting. *Environ. Resour. Econ.* **1999**, *13*, 473–491. [CrossRef]
62. Knussen, C.; Yule, F.; MacKenzie, J.; Wells, M. An analysis of intentions to recycle household waste: The roles of past behaviour, perceived habit, and perceived lack of facilities. *J. Environ. Psychol.* **2004**, *24*, 237–246. [CrossRef]
63. Saphores, J.-D.M.; Nixon, H.; Ogunseitan, O.A.; Shapiro, A.A. Household willingness to recycle electronic waste. *Environ. Behav.* **2006**, *38*, 183–208. [CrossRef]
64. Chan, K. Mass communication and pro-environmental behaviour: Waste recycling in Hong Kong. *J. Environ. Manag.* **1998**, *52*, 317–325. [CrossRef]
65. Berglund, C. The assessment of households' recycling costs: The role of personal motives. *Ecol. Econ.* **2006**, *56*, 560–569. [CrossRef]
66. Ajzen, I. The theory of planned behavior. *Organ. Behav. Hum. Decis. Process.* **1991**, *50*, 179–211. [CrossRef]
67. Boldero, J. The prediction of household recycling of newspapers: The role of attitudes, intentions, and situational factors1. *J. Appl. Soc. Psychol.* **1995**, *25*, 440–462. [CrossRef]
68. Passafaro, P.; Livi, S.; Kosic, A. Local norms and the theory of planned behavior: Understanding the effects of spatial proximity on recycling intentions and self-reported behavior. *Front. Psychol.* **2019**, *10*, 744. [CrossRef]
69. Vining, J.; Ebreo, A. What makes a recycler? *Environ. Behav.* **2016**, *22*, 55–73. [CrossRef]
70. Jenkins, R.R.; Martinez, S.A.; Palmer, K.; Podolsky, M.J. The determinants of household recycling: A material-specific analysis of recycling program features and unit pricing. *J. Environ. Econ. Manag.* **2003**, *45*, 294–318. [CrossRef]
71. Fullerton, D.; Kinnaman, T.C. Household responses to pricing garbage by the bag. *Am. Econ. Rev.* **1996**, *86*, 971–984.
72. Oskamp, S.; Harrington, M.J.; Edwards, T.C.; Sherwood, D.L.; Okuda, S.M.; Swanson, D.C. Factors influencing household recycling behavior. *Environ. Behav.* **1991**, *23*, 494–519. [CrossRef]
73. Hornik, J.; Cherian, J.; Madansky, M.; Narayana, C. Determinants of recycling behavior: A synthesis of research results. *J. Socio-Econ.* **1995**, *24*, 105–127. [CrossRef]
74. Schultz, P.W.; Oskamp, S.; Mainieri, T. Who recycles and when? A review of personal and situational factors. *J. Environ. Psychol.* **1995**, *15*, 105–121. [CrossRef]
75. Garfield, E. Historiographic mapping of knowledge domains literature. *J. Inf. Sci.* **2016**, *30*, 119–145. [CrossRef]
76. Hong, S.; Adams, R.M. Household responses to price incentives for recycling: Some further evidence. *Land Econ.* **1999**, *75*, 505. [CrossRef]
77. Hage, O.; Söderholm, P.; Berglund, C. Norms and economic motivation in household recycling: Empirical evidence from Sweden. *Resour. Conserv. Recycl.* **2009**, *53*, 155–165. [CrossRef]
78. Xie, H.; Zhang, Y.; Wu, Z.; Lv, T. A bibliometric analysis on land degradation: Current status, development, and future directions. *Land* **2020**, *9*, 28. [CrossRef]
79. Rajendran, S.; Hodzic, A.; Scelsi, L.; Hayes, S.; Soutis, C.; AlMa'adeed, M.; Kahraman, R. Plastics recycling: Insights into life cycle impact assessment methods. *Plast. Rubber Compos.* **2013**, *42*, 1–10. [CrossRef]
80. Garfield, E.; Sher, I.H. KeyWords Plus™—Algorithmic derivative indexing. *J. Am. Soc. Inf. Sci.* **1993**, *44*, 298–299. [CrossRef]
81. Zhang, J.; Yu, Q.; Zheng, F.; Long, C.; Lu, Z.; Duan, Z. Comparing keywords plus of WOS and author keywords: A case study of patient adherence research. *J. Assoc. Inf. Sci. Technol.* **2016**, *67*, 967–972. [CrossRef]
82. Mori, Y.; Kuroda, M.; Makino, N. *Multiple Correspondence Analysis*; Springer: Singapore, 2016; pp. 21–28. [CrossRef]
83. Cobo, M.J.; López-Herrera, A.G.; Herrera-Viedma, E.; Herrera, F. An approach for detecting, quantifying, and visualizing the evolution of a research field: A practical application to the Fuzzy Sets Theory field. *J. Informetr.* **2011**, *5*, 146–166. [CrossRef]
84. Fernández-González, J.M.; Díaz-López, C.; Martín-Pascual, J.; Zamorano, M. Recycling organic fraction of municipal solid waste: Systematic literature review and bibliometric analysis of research trends. *Sustainability* **2020**, *12*, 4798. [CrossRef]
85. Camón Luis, E.; Celma, D. Circular economy. A review and bibliometric analysis. *Sustainability* **2020**, *12*, 6381. [CrossRef]
86. Cahlik, T. Comparison of the maps of science. *Scientometrics* **2000**, *49*, 373–387. [CrossRef]

87. Callon, M.; Courtial, J.P.; Laville, F. Co-word analysis as a tool for describing the network of interactions between basic and technological research: The case of polymer chemsitry. *Scientometrics* **1991**, *22*, 155–205. [CrossRef]
88. Soundararajan, K.; Ho, H.K.; Su, B. Sankey diagram framework for energy and exergy flows. *Appl. Energy* **2014**, *136*, 1035–1042. [CrossRef]

sustainability

MDPI

Article

A Systematic Technique to Prioritization of Biodiversity Conservation Approaches in Nigeria

Valentine E. Nnadi [1,*], Christian N. Madu [1,2] and Ikenna C. Ezeasor [1]

[1] Shell Centre for Environmental Management & Control, University of Nigeria, Enugu Campus, Enugu 400241, Nigeria; cn_madu@yahoo.com (C.N.M.); ikenna.ezeasor@unn.edu.ng (I.C.E.)

[2] Department of Management and Management Science, Lubin School of Business, Pace University, 1 Pace Plaza, New York, NY 10038, USA

* Correspondence: ejour2020@gmail.com or valentine.nnadi@unn.edu.ng

Abstract: There are generally no acceptable views on the conservation of biodiversity because there are no known best approaches to that. This has presented a challenge on what and how to conserve in developing countries like Nigeria. This paper used a multi-criteria decision-making model based on the Analytic Hierarchy Process (AHP) to elicit experts' opinions on biodiversity conservation approaches and their corresponding conservation targets. The rationality of the experts was checked by measuring their consistency in the decision-making process. A greedy search algorithm based on linear programming application was also used for resource allocation. This technique is holistic and allows the decision maker to consider all pertinent factors. The approach allows policy makers to integrate worldviews; culture; diverse flexibility of concerned communities and other stakeholders in identifying conservation practices to achieve sustainability. In terms of current performance for the biodiversity conservation approaches; the conservation experts rated their performance on Ecosystem-service-based approach high with the priority index of 0.460. Their performances on Area- and Species-based approaches are ranked second and third with priority indexes of 0.288 and 0.252 respectively. Conversely; in the case of expectations; Ecosystem service is the most important with a priority index of 0.438 followed by Area-based with a priority index of 0.353 and Species–based with a priority index of 0.209. The Ecosystem-service based approach has the highest contribution coefficient. Resources are allocated accordingly; in form of capacity building; based on the priorities that were obtained. The research is a rights-based tool for capacity building; and a paradigm shift from the purely scientific approach to decision-making. It is designed to bridge a scientific gap between policy formulation and resource allocation in biodiversity conservation.

Keywords: analytic hierarchy process (AHP); ecosystem management; expert opinion; environmental planning and modeling; rights-based tool; multi-criteria decision making; optimisation; sustainability

Citation: Nnadi, V.E.; Madu, C.N.; Ezeasor, I.C. A Systematic Technique to Prioritization of Biodiversity Conservation Approaches in Nigeria. *Sustainability* **2021**, *13*, 9161. https://doi.org/10.3390/su13169161

Academic Editor: Alan Randall

Received: 8 July 2021
Accepted: 13 August 2021
Published: 16 August 2021

Publisher's Note: MDPI stays neutral with regard to jurisdictional claims in published maps and institutional affiliations.

1. Introduction

Ecosystem dehydration may lead to the endangerment of some species [1–5]. Several studies have identified both social and ecological criteria to evaluate biodiversity [6–12]. Emphasis has been made in identifying the important criteria for approaches used in the conservation of biodiversity [13,14]. Many studies have also explained the way a number of approaches can be used to examine the social preferences of stakeholders for various situations involving biodiversity and the natural ecosystems [15–21]. For instance, in identifying different stakeholder perceptions of diverse ecosystem services using an ecosystem service-based approach, local actors preferred drinking water, fresh air and climate change control, genetic pool of plant communities and educational value [15].

However, no mention was made of prioritizing the different conservation approaches and their targets by considering the multiple criteria that may be involved. Priorities are sometimes established for conservation at the species level [22,23] but not for the evaluation of the conservation approaches. Similarly, priorities are also often set for hotspots, representing transitions, but not for all targets of biodiversity conservation [24–26]. Despite the existence of many research studies on biodiversity conservations, only a few emphasize the modeling aspect [27,28]. The majority of the studies are conceptual. The few models as noted by [29] are not complete in terms of application because they do not present a consolidative strategy and seem to overstress the significance of the economic and ecological aspects of biodiversity conservation. These articles seldom emphasize the conservation subsystems—the targets. Furthermore, no effort is made to put together these subsystems to form a general framework for modeling biodiversity conservation approaches. The shortcomings of the noted models led [29] to apply Analytic Hierarchy Process (AHP) as a model for making choice on biodiversity preserved areas. However, the author did not investigate the biodiversity conservation approaches (BCAs) and their conservation targets (CTs).

This paper extends the work of [29] by developing a ranking system for BCAs and the CTs. This is done through the application of AHP to develop priority indices for both BCAs and the corresponding CTs. This paper further used an input–output model. As a result, the concept of "mutual dependence" amongst the different units of the conservation approaches is introduced and used in building a model for limited resource allocation strategies. By mutual dependence, we mean that some approaches may yield more benefits if there are other approaches in existence that may serve as supports for their activities. For instance, output from area-based conservation may be an input for species-based conservation and vice versa, or an output for species-based conservation may be an input for ecosystem-service based conservation. This situation may or may not be a two-way process. Nevertheless, the reality of such mutual dependence between or amongst different approach types may present synergistic benefits.

Every nation faces the issue of limited resources, especially developing countries, in terms of funding. There are many social services demanding financial support. Therefore, there is a need to optimize the nation's limited resources and further cascade down to all the units to ensure that these resources are properly utilized to address the most important needs. Ecological management is a forefront issue in Nigeria as the country aims to sustain increasing population, unemployment and other social issues. The long-term goal of building a biodiversity conservation capacity to effectively support Nigeria's economy may be attained through the effective prioritization of BCAs and their CTs using expert opinion. A functional procedure for the allocation of Nigeria's limited resources is developed. The process recognizes all major factors considered by the experts.

This paper presents a balanced approach for making decisions that permit a holistic consideration of biodiversity conservation opportunities and the final creation of a new dimension of conservation approach capable of improving prospects for conservation in Nigeria. The study is based on measuring the perceptions of biodiversity conservation experts. Madu and Madu [30] described such group decision-making used to gain acceptance of the decisions made as a *bargaining window*. Even though we focus on applications to Nigeria's biodiversity conservation management, the approach followed here can be widely applied in other developing countries where data collection is usually rough and may not be available in certain cases. Thus, the emphasis in Nigeria is illustrative but the modeling approach presented here is the major contribution of the paper. The research is a rights-based tool for capacity building, and it is designed to bridge a scientific gap between policy formulation and resource allocation in biodiversity conservation. It supports the 15th goal of the 2030 UN agenda for Sustainable Development, designed to halt biodiversity loss.

2. Methodology

Analytic Hierarchy Process (AHP)

This study used a Multi-Criteria Decision-Making (MCDM) technique. The technique requires the use of experts to consider multiple criteria that may affect biodiversity conservation decisions. Specifically, an Analytic Hierarchy Process (AHP) is applied here to prioritize the approaches and targets of biodiversity conservation. In Figure 1 below, the biodiversity conservation approaches are displayed along with the conservation targets for evaluation using AHP. What this implies to the Area-based approach for instance is that, in its target, it is either that the area is preserved because it is endemic or that it is conserved to protect a non-endemic area.

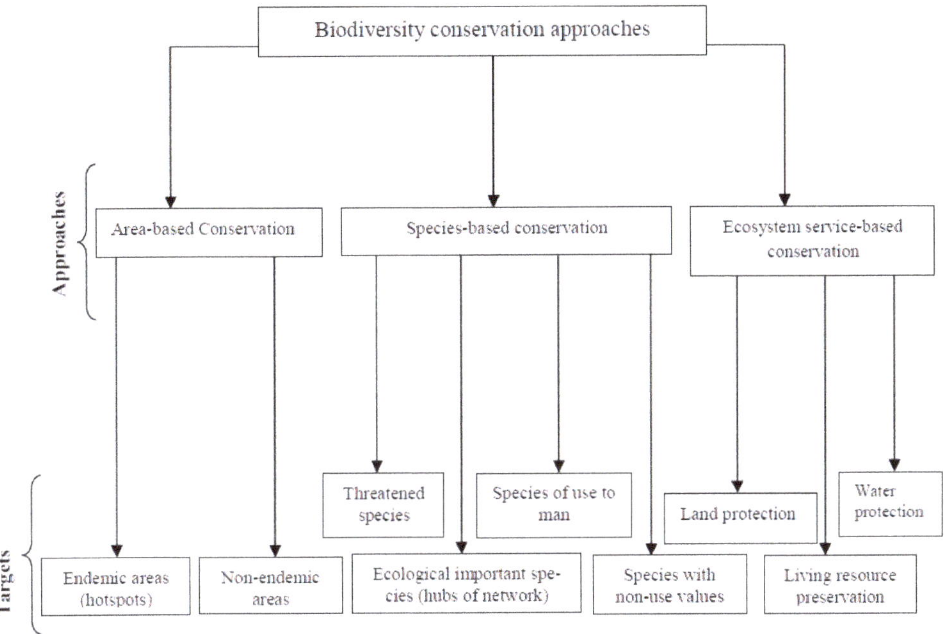

Figure 1. Biodiversity conservation approaches (BCAs) and their corresponding specific conservation targets (CTs).

The AHP is a special form of MCDM that is used here and has a wide range of applications in different disciplines [31–36]. The multi-criteria decision-making structure based on AHP is presented in Figure 2 below and integrated as steps 2–9 in Figure 3. This structure provides an opportunity for stakeholder/expert judgments for the choice of biodiversity conservation approaches (BCAs) and conservation targets (CTs). Specifically, the steps represented in blocks and numbered in the figures are discussed briefly in the ensuing discussions.

First, the experts are identified (Figure 2). Policy makers in Nigeria participate actively in decisions to conserve biodiversity. Their active participation may include making decisions on the right approaches to conserve biodiversity, the allocation of resources to improve biodiversity conservation, socioeconomic, legal and ecological management and others. As a result of these, the structure of Figure 3 starts with the policy maker (1) as the initiator of conservation policies. It is presumed that the policy maker is aware of the importance of biodiversity conservation as a major part of ecosystem balancing and intends to make it an integral part of ecosystem management plan.

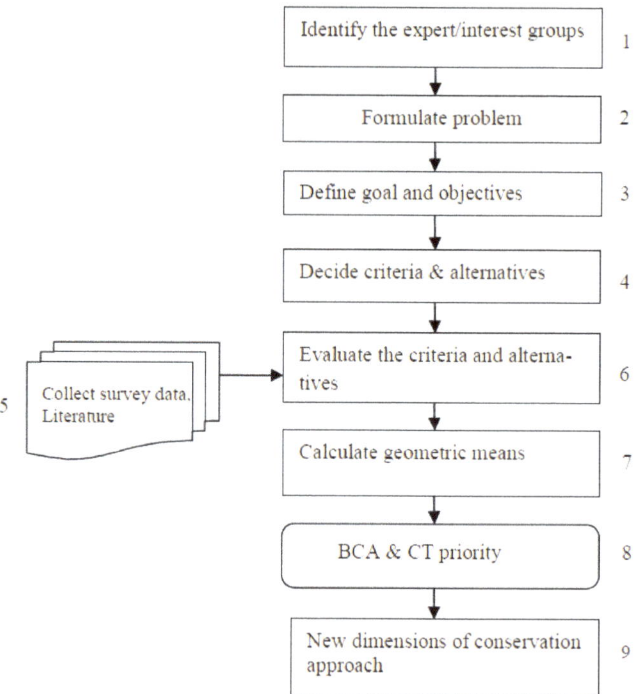

Figure 2. Methodology of analytical hierarchy process (AHP) for prioritizing biodiversity conservation approaches (BCAs) and conservation targets (CTs).

There are several interest groups and individuals who may partake in such decisions or policy making. Consequently, in Figure 3, the policy maker identifies interest and/or expert groups (2) whose ideas, opinions and perceptions may affect the decisions related to the successful conservation of biodiversity. The interest groups/experts here comprise the local people who are working with Biodiversity Conservation Community-Based Organisations (BC-CBOs). By involving this set of people, local or indigenous knowledge is incorporated. Experts are also drawn from non-governmental organizations (NGOs) and government agencies and parastatals whose primary objectives are on biodiversity conservation.

In step 3 of Figure 3 which is the second stage in Figure 2, the conservation problem is formulated as a "mess" [37]. The "mess" here is the case of depleting biodiversity against all efforts to conserve it. Step 4 of Figure 3, which is the third stage of Figure 2, suggests defining the goals, objectives and sub-objectives. Immediately the objectives are defined, the strengths and weaknesses are outlined. Subsequently, a set of criteria are identified to achieve these objectives in line with [31,32]. Consequently, step 5 of Figure 3 (the fourth stage of Figure 2) calls for deciding and/or evaluating the performance criteria and alternatives for each major objective. Marcot et al. [38] define objectives as "the long-range aspirations of the decision makers and stakeholders," and can include ecological, economic, recreational, spiritual, cultural and aesthetic dimensions in the case of biodiversity conservation. Primary objectives may oftentimes be structured into hierarchies, as can be observed in the decision alternatives of biodiversity conservation. For instance, there may be several ecological and economic objectives for biodiversity conservation plans, with some potentially conflicting with each other. Organizing these fundamental objectives in a hierarchical order can be helpful in explaining tradeoffs and priorities amongst many objectives.

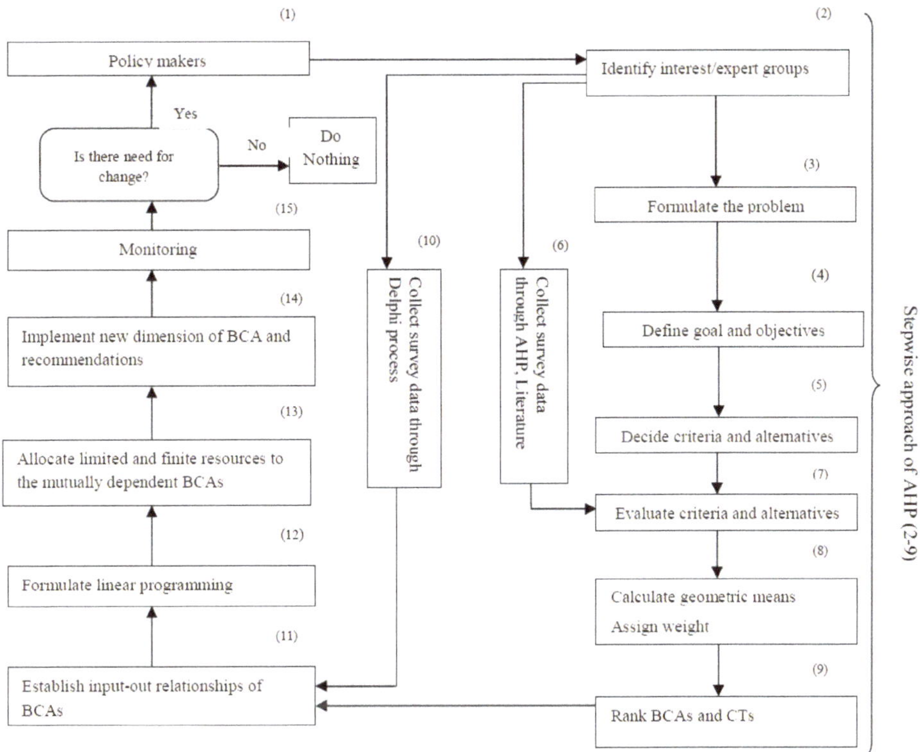

Figure 3. A systematic method to improving biodiversity conservation approach.

In Figure 4 below, level 1 is the goal of the project, which is to improve biodiversity conservation. There are certain criteria established at level 2, in line with the demands of step 5 of Figure 3 to achieve this set goal. Many variables such as species populations, species traits, community composition, ecosystem structure and ecosystem function have been used in quantifying these criteria [39,40], though they differ across various biodiversity conservation initiatives. However, to the best of our knowledge, no single study integrates all of these criteria. First, we reviewed the ecological and economic criteria applied in various initiatives to determine the main criteria common in most initiatives. Second, we then synthesized the biodiversity conservation variables needed to inform these criteria. These criteria give direction on the direct assessment of approaches important for biodiversity conservation as well as the conservation targets. These criteria are social, economic, environment, cultural, research and development, resource utilization and the cost of conservation. These are constraints to the decision variables or decision alternatives at level 3 in Figure 4. In other words, the decision alternatives ought to be prioritized considering the criteria to achieve the set goal. Deciding the decision alternatives includes first identifying exact decision variables (the items that are controllable) and the ranges that are acceptable for the variables (e.g., species- or ecosystem service-based approach) and, second, generating alternatives based on those variables. Here, a list of the conservation projects (species-based, ecosystem service-based and area-based approaches) as well as the conservation targets form the alternative portfolio for the domain experts to compare and rank. However, the caveat of this study is the absence of landscape scale in the biodiversity conservation approaches considered. This could be an important area to consider for future research.

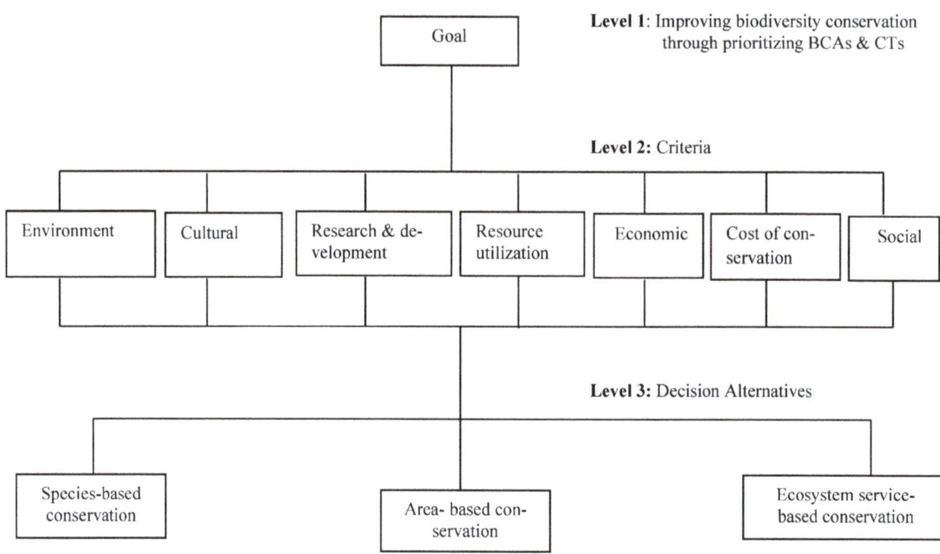

Figure 4. A network structure for prioritizing biodiversity conservation approaches (BCAs).

Step 6 of Figure 3 and stage 5 of Figure 2 suggest survey data collection using AHP. In this context, the survey data are the information from the domain conservation experts. For the application of the AHP in this study, a survey research design that is based on pairwise comparison of the decision makers' judgment is used. The survey instrument has three sections, namely, A, B and C. Section A is designed to elicit the demographic data of the experts. In sections B and C, a 9-point scale is used to conduct pairwise comparisons on the conservation approaches and conservation targets of each of the approaches, respectively, for the expectation or capacity building. The 9-point scales are defined as follows: 1 = equal importance; 3 = moderate importance; 5 = strong importance; 7 = very strong importance; 9 = extreme importance of one action over the other. The even numbers 2, 4, 6, and 8 are used for compromise while reciprocals are used to show inverse comparisons. This is the standard scale that is normally used with AHP. However, alternative scales could be used. Conservation experts ascertained the comprehensiveness of the top-level conservation approaches and their targets. Through the AHP, a sequence of pairwise comparison is performed between alternative actions or decisions (i.e., biodiversity conservation approaches and the conservation targets). A purposive sampling technique was used in selecting 28 biodiversity conservation organizations that participated in the study. This sampling technique is effective when studying a cultural domain with knowledgeable experts within [41]. The organizations have people who work directly on biodiversity conservation and our interest is to tap into their expert knowledge. Selection of participants for this study is based on the specialized education, knowledge, experience, skill and training acquired that qualified them as domain experts.

The pairwise comparison survey instrument was administered to two biodiversity conservation experts in each of the 28 organizations. Thus, 56 pairwise comparison survey instruments were distributed. The returned instruments were evaluated using the AHP. This process requires several iterations that demand back-and-forth communication with the experts to address areas of inconsistent judgment. Some of the experts did not follow through with some of the iterations and, therefore, did not complete the process. As a result, they were dropped. Thus, 27 experts were dropped along the line, and only 29 experts completed the iteration processes. This number is significant in group decision making since group decision making does not require large samples to avoid potential conflict and achieve timely decisions. Furthermore, in using expert judgment, representation is measured by the quality of the experts and not by the numerical size [42]. Several research works have adopted similar approach [1,31,32,43–52]. Therefore, this research design is proper for the intent of this study. The AHP technique is used in this context because it allows us to measure the consistency of the experts' judgment, which is a function of their rationality. The rationality of the experts was checked by measuring their consistency, through the AHP, and a Critical or Consistency Ratio (CR) < 0.10 shows consistency in judgment.

Steps 7 and 8 of Figures 2 and 3, respectively, suggest we calculate the geometric means. The method of the AHP was used to analyze the matrices. For ease of analysis, the independent matrices are unified into one matrix by finding the geometric means of each of the cells in the matrices, which are the objects for further analysis or AHP application (see Appendix A). Aczel and Saaty [53] explain that the use of the geometric mean is essential to conserve the reciprocal property of the objects. Stage 8 of Figure 2 corresponds to block (9) of Figure 3 and suggests ranking of the BCAs and CTs. Here, the priority indices are computed for each of the cells in the matrices and ranked using the values of the priority indices. The rank order portrays the order of importance attached on the criteria and alternatives [54]. The ranking of the criteria and alternatives of the BCA is performed to ensure that importance or preference is placed on the most important BCAs and CTs.

Given that the data required to set up the input–output relationship of the BCAs may not be solely quantitative, the Delphi technique is used to gather the data (see Block (10) of Figure 3). The Delphi technique is premised on the principle that decisions made in a structured environment is more accurate than unstructured ones [55]. In this paper, a modified Delphi technique is applied such that a questionnaire is designed with a 5-point linear Likert scale of: 1 = not at all important; 2 = slightly important; 3 = moderately important; 4 = very important; 5 = absolutely important on perceived importance of the interdependence between a given pair of the BCAs. Expert participants are also drawn from the government organizations, Biodiversity Conservation Community-Based Organizations (BC-CBOs) and other NGOs. The questionnaire was shared to the expert participants for rating using the Likert scale, after having the results of the AHP questionnaire. Each of the participants performed the task anonymously to keep away from group behavioral influences. They were asked to review their weight assignments, taking into consideration the reasons stated by other participants so they can justify their original weight assignments. Three iterations were made in the process, and we observed some convergence of opinions. Geometric means were calculated to reflect the consensus of the group in accordance with the guideline provided by [56,57]. The data from the Delphi process suggest that there is tendency that each of the BCAs may be feeding on the other. There is therefore need to explore the mutual dependence between the BCAs.

Block (11) of Figure 3 suggests establishing input–output relationships. The relationships are very important since resource allocation cannot be based mainly on priorities when there may be interdependencies existing. This means that there is need for optimization in the face of limited resources, and therefore, there is need to introduce Leontief input–output model. The input–output table (matrix) developed from the data elicited through Delphi is then used with the priority indices on the expectations of the three BCAs for application of the Leontief input–output model (see Appendix B). Again, block (12) of Figure 3 suggests the formulation of a linear programming (LP) problem (see Appendix C) since the efficient utilization of resources is a requirement to achieving the targets of the BCAs in environmental planning and management.

3. Results and Discussion

Using the AHP, the priority indices of the BCAs are computed (see Appendix A). The pairwise comparison matrices of the three BCAs and their corresponding CTs are based on the relative importance ratings of both their expectations and actual performances. The aim is to generate data for prioritizing the BCAs and the CTs to see how we are faring and perhaps identify areas for capacity building and to efficiently allocate limited resources.

3.1. Prioritizing the Conservation Approaches

In the case of expectations for the three biodiversity conservation approaches, as shown in Table 1, the ecosystem-service-based approach is perceived to be the most important with a priority index of 0.438, followed by the area-based approach with a priority index of 0.353, and the species-based approach ranked third with a priority index of 0.209.

Table 1. Prioritizing the biodiversity conservation approaches (BCA)—Expectation vs. actual performance.

Conservation (Top-Level) Approaches	Expectation		Performance	
	Priority Index	**Ranking**	**Priority Index**	**Ranking**
Species-based	0.209	3	0.252	3
Area-based	0.353	2	0.288	2
Ecosystem-service-based	0.438	1	0.460	1

CR = 0.094.

However, in terms of current performance as also shown in Table 1, it is clear that the respondents rated their performance on ecosystem-service-based approaches high, as is evident in the priority index of 0.460. Their performances on area-based and species-based approaches are ranked second and third, respectively, with priority indices of 0.288 and 0.252, respectively. Furthermore, the Consistency Ratio (CR) of 0.094 shows that the experts are consistent in reaching this conclusion.

The technique we used permits us to rank, in order of priorities, the three BCAs and the CTs related to each of the approaches. The generated priority indices of the three BCAs (BCA1, BCA2 and BCA3) suggest that there is varying perceived importance of each approach in contributing to biodiversity conservation. The matrix implies that BCA3 and BCA2, with priority indices of 0.438 and 0.353, respectively, should have the highest priorities. On the other hand, BCA1 seems to have the least preference based on all the criteria considered. Though we underscore the importance of all the approaches, their rank order of importance is relative, and to achieve the goal of biodiversity conservation, emphasis should be on the approaches with the most perceived importance. Conversely, in terms of actual performance, the ecosystem-service-based approach (BCA3) is perceived to be the most important. The perceived importance is in line with the preference given to the approach in the MEA report [58]. The participants' reason for placing higher preference on BCA3 may be that the approach cuts across all sections of sustainability (economic, social and environment) and so it is holistic. This agrees with the reports of [59–63]. We also seem to do relatively well in each of the other two conservation approaches, that is, the area-based and species-based approaches. This may suggest a wide coverage on the different approaches. However, the lower rank order of BCA1 and BCA2 may be linked to emphasis being shifted from these approaches after the UN convention on biological diversity (CBD) that showed preference to the ecosystem-services approach (CBA3). Our results are consistent with the reports of the [64,65]. It is also noteworthy that since the expectations of the area-based approach are high in importance but low in the case of performance, it suggests that improvement through capacity building and provision of resources to support the approaches is required. The result shows that the rank order for the conservation approaches in terms of actual performance coincides with the rank order for expectations. In the area-based approach, the perceived importance is higher in expectation than in current performance. However, the perceived importance of expectations is slightly lower in the species-based and ecosystem-service-based approaches in comparison with their current performance. This may partly explain the institutional development or local/city governance issues. The results of the pairwise comparison matrices show that a gap exists between the perceived expectation and the actual or current performance.

3.2. Prioritizing Conservation Targets under the Species-Based Approach

In Table 2, the rank order for both expectation and performance is preserved for the targets under the species-based approach. However, as seen in Figure 5, the perceived importance to build capacity (expectations) on the targets is slightly lower in CT2 (Ecological important Species) whereas CT4 (Species of non-use to human) is slightly higher when compared to their priority indices in the current performance. Again, in capacity building, CT3 (Species of use to human) completely dominates CT1 (Threatened Species), CT2 (Ecological important Species) and CT4 (Species of non-use to human). CT4 is of the least perceived importance here, lower than CT2 and CT1, which are ranked second and third, respectively, though their margin is very slim.

Table 2. Prioritizing the conservation targets under the species-based approach—Expectation vs. actual performance.

Conservation Target (CT)	Indicators	Expectation		Performance	
		Priority Index	Ranking	Priority Index	Ranking
CT1	Threatened species	0.190	3	0.190	3
CT2	Ecological important Species	0.194	2	0.209	2
CT3	Species of use to human	0.435	1	0.393	1
CT4	Species of non-use to human	0.182	4	0.175	4

CR = 0.074.

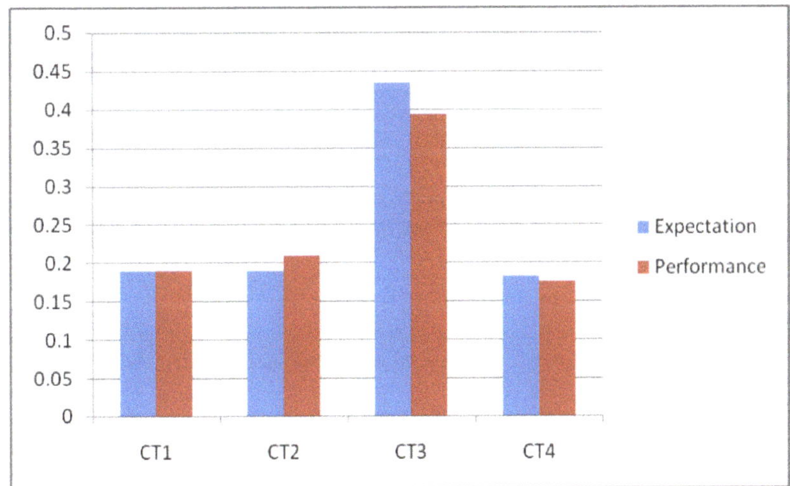

Figure 5. Chart of the perceived performance and expectation of the species-based approach (BCA1).

Though the rank order is maintained in the case of current performance as it is in expectation, there is variation in the priority indices of the CTs, except for CT1 (Threatened species), which is exactly the same in both expectations and actual performance, as shown in Table 2. In terms of current performance, the priority index of CT3 (0.393) is higher than that of CT1 (0.190), CT2 (0.209) and CT4 (0.175). The priority index of CT2 is marginally higher than CT1 but clearly above CT4. The experts are consistent in their judgment, as is evident in the CR value of 0.074.

As expected, CT3 (Species of use to human) is perceived to be the most important, and it is significantly higher in preference than the other targets, both in expectations and in actual performance. CT4 (species of non-use to man) as expected is the least in both expectations and actual performance. However, the participants perceive that more capacity needs to be built on CT4 probably because of the need for future use. CT2 (Ecological important Species) is ranked second, after CT3, in both expectations and actual performance. This may be because of the fact that any species of ecological importance promote environmental sustainability and, as such, partly contributes to the eco-efficiency and effectiveness. The reason may not be far from why CT2 is referred to as the "hubs of network". CT1 (Threatened species) is ranked third probably because Nigeria's biodiversity richness may be protective of many of these species. Furthermore, efficient and effective hubs of networks (CT2) may also protect threatened species.

3.3. Prioritizing Conservation Targets under the Area-Based Approach

In order of importance in both expectations and actual performance, the area-based conservation approach is ranked second over the species-based approach. There are two set targets to this approach: CT5 (Endemic area—hotspot) and CT6 (Non-endemic areas) [29]. In terms of capacity building, we notice that the CT5 is of higher perceived importance than CT6 for the targets to be achieved (Table 3). This supports the work of [66]. However, in terms of current performance, there is no dominance between the targets (i.e., CT5 and CT6). They are of equal ranking with priority indices of 0.500 each. Meanwhile, as expected, the rank order of CT5 is higher in expectations than in actual performance, but the perceived importance of CT6 is higher, with a priority index of 0.500 over that of expectations with 0.394. There is consistency in the judgment of the experts since the CR value is 0.031.

Table 3. Prioritizing the conservation targets under the area-based approach—Expectation vs. actual performance.

Conservation Target (CT)	Indicators	Expectation		Performance	
		Priority Index	Ranking	Priority Index	Ranking
CT5	Endemic areas (hotspot)	0.606	1	0.500	1
CT6	Non-endemic areas	0.394	2	0.500	1

CR = 0.031.

As also expected, the CT5 (Endemic areas), referred to as "hotspots", are perceived higher than the CT6 (non-endemic areas) in terms of expectation, probably because endemic areas are losing biodiversity to the built environment. This may be because of our poor institutional development or city/local governance where personal interests override sustainable development goals, as reported by [67]. The agreement on the perceived importance in terms of actual performance may perhaps be due in part to their struggles to develop non-endemic areas, which create balance for the lost endemic areas. However, the experts may choose to have equal preference in CT5 and CT6 probably because conservation of any one of them is not guaranteed by the current public policies. Situations arise where the public sector does not consider endemic or protected areas over non-endemic areas. Infrastructural developments are often approved to the detriment of the endemic areas.

3.4. Prioritizing Conservation Targets under the Ecosystem-Service-Based Approach

The ecosystem-service-based approach is the first in terms of expected priority ranking and even in terms of actual performance. In other words, it is perceived to be of most importance of all in both expectations and current performance. The approach sets three targets: protection of water bodies (CT7), protection of land (CT8) and protection/preservation of living resources (CT9) [15,21], as shown in Table 4. In terms of actual performance, CT7 (water) is ranked to have low performance, with a priority index of 0.121. Therefore, there may be need for capacity building here. CT9 (Living resources) is ranked first, as is evident from the high priority index of 0.268 above CT8 (land) and CT7 (water) with priority indices of 0.247 and 0.121, respectively. The result of our evaluation shows that the priority indices generated for each of these CTs are relatively close to each other, thus suggesting the perceived importance of each in effective use of this approach.

Table 4. Prioritizing the conservation targets under the ecosystem-service-based approach—Expectation vs. actual performance.

		Expectation		Performance	
Conservation Target (CT)	**Indicators**	**Priority Index**	**Ranking**	**Priority Index**	**Ranking**
CT7	Water	0.204	3	0.121	3
CT8	Land	0.251	2	0.247	2
CT9	Living Resources	0.273	1	0.268	1

CR = 0.083.

In the case of expectations in Table 4 above, CT9 is ranked first with a priority index of 0.273 above CT8 (0.251) and CT7 (0.204). CT7 is ranked third with a priority index of 0.204. The CR value of 0.083 shows that the experts are consistent in reaching this conclusion.

More so, the CT7 (Water), CT8 (Land) and CT9 (Living resources) share the same rank order (first, second and third, respectively) in terms of both expectations and actual performance and show that much is actually being done. This is evidenced in the dredging of some of the waterways and sensitization on the dangers of dumping wastes into the available rivers, the clean-up of oil spill sites, erosion control and other sustainability activities. However, a gap exists which shows that much is also expected or that there is need for capacity building in that order. The rank order, in terms of performance on CT9 (preserving the living resources), may be due to the unsustainable pattern of consumption in the country. This may perhaps explain the need to do more in using area- and species-based conservation approaches as stated earlier. Again, the low ranking of CT7 (preserving or protecting the water bodies) in terms of actual performance may be due to the negative effects of some anthropogenic activities in Nigeria. For instance, the common practice or attitude of dumping wastes or other hazardous substances into the water bodies often contribute to the massive flooding that is experienced in the country. Even though we do well in CT8 (land protection) as shown in Table 4, expectations are high and require that capacity needs to be built in protecting land to enjoy the land-related benefits of BC3. Although some of these processes such as land restoration are natural phenomena, without taking care of the land, the ecosystem service benefits may not be realized. Expectation through capacity building on the ecosystem-service-based approach is required, as shown in the ratings. It is observed that although the ratings differ amongst the three CTs in terms of actual performance, their priority indices are relatively close and may suggest same level of perceived performance.

Generally, it is expected that high rank order conservation targets should consume more resources. However, considering the holistic process of BCA3 (ecosystem-service-based approach) from a management point of view, it may be better to *satisfice* rather than *optimize*. In other words, rather than distribute the resources disproportionately to the targets with significantly higher rank order, it may be more preferable to allocate such resources more equitably to encourage the achievement of all set targets and/or benefits in the approach.

In fact, we must emphasize that the rank order of the majority of the expectations of the BCAs and related CTs coincides with their respective actual performances. In other words, there seems to be some conformity in the ranking of some BCAs and CTs in both cases. However, there is notable difference in their priority indices. Consequently, we can stress that virtually all the BCAs and related CTs are important. Their order of importance is relative. Attention should therefore be given to the BCAs with the highest perceived importance but also given equitably to CTs. The rating of the current performance facilitates the identification of the gaps in deployment and underscores what we may not be doing at present. The expected importance rating provides vital information because it demonstrates in a rank order where interest should be channeled to, to achieve robust biodiversity conservation.

3.5. The Input–Output Relationships

Block (11) of Figure 3 recommends establishing input–output relationships of the BCAs. Table 5 is the priority indices of expectations in the three top-level biodiversity conservation approaches. Table 6 is an outcome of the established input–output relationship of the three conservation approaches obtained through the Delphi technique. The coefficients in the (i,j) cells are weighted by the α_i (Table 5) and α_j (Table 6), respectively, and a summation is made over each row to obtain the dependence vector matrix β [30,68] given as Table 7. This matrix presents the adjusted weights for all BCAs. The matrix considers both the originally derived expectation priorities in case of the criteria for improving biodiversity conservation with the interdependencies among the conservation approaches. The α_i and α_j are eigenvectors derived for the BCAs.

Table 5. α matrix for the biodiversity conservation approaches (BCAs).

	BCA1	0.209
$\alpha =$	BCA2	0.353
	BCA3	0.438

Table 6. The input–output matrix for the biodiversity conservation approaches (BCAs).

	BCA1	BCA2	BCA3
BCA1	3.60	3.99	3.56
BCA2	3.62	4.54	3.65
BCA3	3.44	3.75	4.24

Table 7. The dependence vector matrix for the biodiversity conservation approaches (BCAs).

	BCA1	0.778
$\beta =$	BCA2	1.397
	BCA3	1.708

The values of the geometric means of the participants' responses, from the Delphi process, fall between 3.60 and 4.54. This implies that the perceived importance of the interdependence between a given pair of the BCAs is either moderately important or very important from the range of Likert scale adopted. This means that there is flow amongst the three conservation approaches (BCA1, BCA2 and BCA3). They exhibit interdependence to an extent. Steps to use one may ease the other showing that they are mutually dependent on one another. This synergistic influence may aid in allocating resources to achieve set targets. Notice also that a BCA may be partly dependent on itself; that is, a conservation approach may partially depend on itself, hence some of its output is retained for internal use to improve the approach itself. The observed interdependence between the three BCAs necessitates the use of linear programming for resource allocation.

The network of interdependence for the three BCAs is shown in Figure 6. The outside straight arrows that connect the spheres in the diagram show the interflow amongst the BCAs. This means that each of the BCAs (BCA1, BCA2 and BCA3) feeds on each other and are therefore mutually dependent. An attempt to use one may enhance the other. Again, the curved right arrows inside the spheres show the internal use of the outputs by each of the BCAs or that some of the outputs are retained internally to keep the system working.

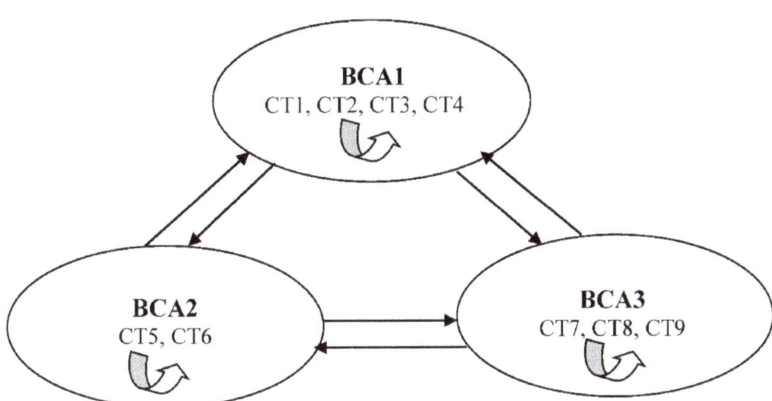

Figure 6. Interdependence network of the three biodiversity conservation approaches (BCAs).

3.6. Allocate Limited and Finite Resources to the Mutually Dependent BCAs

Block (13) of Figure 3 in the methodological framework proposes the allocation of scarce funds to the mutually dependent BCAs. This resource allocation system uses the input–output matrix and the priority indices generated for the different approaches. Table 8 shows the resource needs for these approaches.

Table 8. Sharing of the NGN 100,000,000 budget allocation for the biodiversity conservation approaches (BCAs).

Biodiversity Conservation Approaches (BCAs)	Resource Requirement (Millions of NGN)	Real Matrix (w) (Resource Requirement/Total)
BCA1	39	0.339
BCA2	33	0.287
BCA3	43	0.374
Total	115	1

For purposes of clarity, we found that in 2019 a specific budget of NGN 100 million is committed to accomplishing the goal of biodiversity conservation in Nigeria. Assume that each of the units that cover the three approaches have submitted their resource requirements as shown in Table 8. There must be allocation of appropriate resources in order to consider all identified criteria and their matching priorities in maximizing the country's biodiversity conservation.

Using Equations (A4)–(A6) (see Appendix C), the step followed in [30–32,68], we can develop the linear programming model for the following problem:

$$\text{Maximum } z = 0.778w_1 + 1.397w_2 + 1.708w_3 \tag{1}$$

Subject to

$$0 \leq w_1 \leq 0.39$$
$$0 \leq w_2 \leq 0.33$$
$$0 \leq w_3 \leq 0.43$$

and

$$w_1 + w_2 + w_3 = 1 \tag{2}$$

where the β vector matrix presents the coefficients of the objective function while the upper bounds of the first three constraints are the ratios of the funds needed by each BCAs to the NGN 100,000,000 fund allocation. The decision variables represent the ratio of funds allotted to the three BCAs, as illustrated in Figure 7. The results found here are rather intuitive even though the challenge can be handled by applying specialized LP software package such as LINDO [69], OpenSolver or *R*.

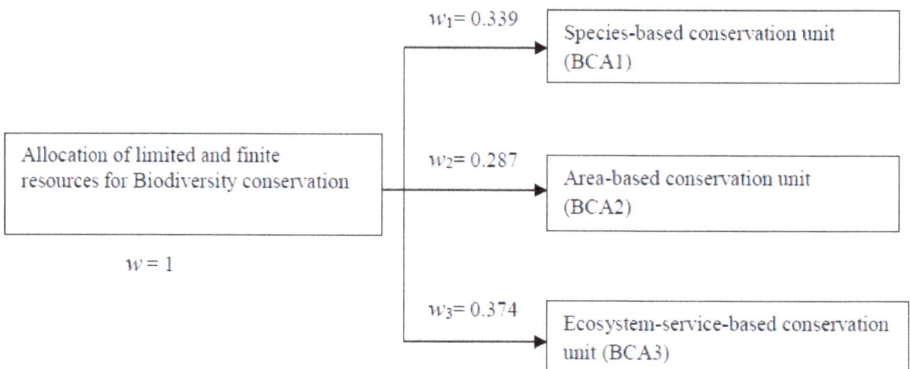

Figure 7. Biodiversity conservation resource allocation priority.

The technique used here is a greedy search algorithm whereby the most important approach gets its maximum requirements in that order of priority and if in the end anything is left, it will be allocated to the least priority approach. In other words, the optimal allocation decision can be made by a simple ranking of the BCAs on the grounds of their β-matrix values (contribution coefficients) and, after that, using the constraints to fund approaches to their needed resource until the fund is depleted. For instance, from the coefficients of the objective function *z*, we find that BCA3 has the highest contribution coefficient of 1.708. Thus, w_3 = 0.43, meaning that BCA3 will receive its complete resource need. Given that we have a balance of $(1 - 0.43 = 0.57)$, we have enough to meet the complete resource needs of BCA2. Again, 1.397 for BCA2 is the next highest contribution coefficient, implying that w_2 = 0.33.

Note that the last constraint in the model is $w_1 + w_2 + w_3 = 1$. Therefore, $w_2 + w_3 = 0.76$. The next conservation approach is BCA1 with the least contribution coefficient of 0.778. This presents w_1 to be assigned $w_1 = 1 - 0.76 = 0.24$. It is observed that all approaches with the exception of BCA1 received their full allocation of funds. The balance of 0.24 is less than the 0.39, which is the proportion of need for BCA1. Therefore, we do not have sufficient resources to contain that need. Generally, this analysis is called "greedy heuristic" because the allocation of resources to the approaches follows a non-increasing order of the contribution coefficients until all the resources are exhausted. For this problem, the resource allocation based on the greedy heuristic algorithm is shown in Table 9.

Table 9. Resource allocation based on the "greedy heuristic" algorithm for the biodiversity conservation approaches (BCAs).

Biodiversity Conservation Approaches (BCAs)	Resource Allocation (Millions of NGN)
BCA1	24
BCA2	33
BCA3	43

3.7. Implement New Dimension of BCA and Recommendations

Block (14) of Figure 3 suggests the implementation of new dimensions of BCA. Successful implementation of a decision in biodiversity conservation involves taking into account the risks and benefits, time and cost related to the implementation. In this case, the time and cost of implementing this new dimension of biodiversity conservation approach ought to be considered alongside the risks involved, if any, and the benefits of the implementation. The step also considers the feasibility of the developed framework and the level of impact of the implementation to avoid failure. It is expected that the decisions will have the support of considerable interest groups that are represented by the expert respondents since the decisions are founded on the converging opinions of the stakeholders. The results, in addition, provide a negotiating tool to the policy maker, who may possibly have to give reasons to different groups why a particular conservation approach was emphasized on importance and why limited resources should be channeled appropriately.

3.8. Monitor the Implementation Process

Block (15) of Figure 3 emphasizes the importance of continual monitoring of the prioritized BCAs to achieve corresponding targets. The continual monitoring process is a kind of control designed to make sure that Nigeria's commitment to conservation of biodiversity is being fulfilled. Monitoring design is most useful when it emanates from the decision background. The standards of monitoring and the process for monitoring should be identified based on the information requirements of the decision maker. The framework of Figure 3 suggests a feedback loop to continuously monitor the planning process. For example, if, in the block (15), there is a need to adjust, the policy maker may decide to identify new interest groups to evaluate the problem. These interest groups may follow the step-by-step approach presented in Figure 3. Conversely, if there is no need for change, then nothing needs to be done other than to continue with periodic review.

4. Policy Implications/Suggestions for Policy Recommendations

Several policy implications emerge from this research work and are identified as follows:

(1) When decision/policy makers use the concept and tools of this research, they may be able to make quality decisions in conservation planning and management since all quality decisions are rational but not all rational decisions lead to a quality outcome. The systematic technique can provide the fundamental principles and deal with the complexities of natural resource planning and management in several places.

(2) This study is a paradigm shift from the purely scientific approach to decision making. This approach is instrumental in ensuring that the different worldviews and perceptions including local knowledge are considered in policy formulation.

(3) Often, we adopt international standards and guidelines without evaluating our peculiar situations. Lack of local content may lead to the unsuccessful implementation of such programs. The inclusion of local content is crucial in biodiversity conservation decision making. The use of the stakeholder approach presents an opportunity for all important interest groups to partake in policymaking.

(3a) Stakeholders may have varying views and premises. They are able to share these views and also understand the worldviews of others. It is through these kinds of teams that conflicts could be resolved and made productive.

(3b) Adoption of the stakeholders' recommendations may gather support for implementation. This may also be helpful in terms of resource allocation as members of the team would likely defend the decisions which they participated in making.

(4) Since the United Nations has identified biological diversity loss as a worldwide issue to tackle, it is very important to prioritize the approach to meet set goals at both the national and global level. In other words, needful approaches that best suits several circumstances in different geographical regions under different environmental challenges should be adopted in cognizance of global needs.

(5) The BCAs involve multiple objectives that need to be established as set targets. Once a country ranks its actual performance low in any of the BCAs it perceives to be critical to preserve biodiversity loss, it is suggested that there is need for new strategies or to realign existing ones, reallocate resources and possibly build capacities that may be required to achieve the set targets.

(6) Many developing countries such as Nigeria are at the risk of losing their biological diversity. Some of these countries may not have the capacity to conserve biodiversity at any level, such as species, areas or ecosystem, perhaps because of their consumption patterns. Many of them also may not have the economic capacity to conserve and may depend on foreign donors for funding. Therefore, we need to make a choice based on scale of preference or perceived importance. It is imperative to methodically allocate the limited resources so that key targets can be met.

(7) This study presents a rational decision-making process to prioritizing BCAs. It surely may not deal with all important problems or assure an optimum solution however, it may present a consolidative, systems perspective of this pertinent problem.

5. Conclusions

The method followed here presents a logical technique to ecosystem planning and management by putting together all factors that are considered significant and influential in the prioritization of biodiversity conservation approaches. Though the interest groups suggest the implementation of their decision, it is not assured that the policy maker will do so. It does not seem unusual, for instance, for policy makers to go by their instincts; however, at the very least, the process used here guarantees that most of the issues concerning the question of biodiversity conservation will be effectively considered. Models as used here serve as decision supports only.

This research work deals with four major challenges to advancing the BCA for coming years: stakeholder inclusiveness, capacity building, resource allocation and local content adaptation. We set out to deal with these to form a baseline, in both future research and practice, in finding a solution to effectively conserve biodiversity and halt its loss. Nigeria is used as a case study so that we can understand the challenges. Considerable changes to the robust ecosystem and being proactive against biodiversity loss can be realized through the application of the processes and techniques present in this work. We found the process of group elicitation to address conservation planning highly effective. The members of the group could retrospectively provide the logic and reasoning responsible for developing the criteria ranks because a formal decision-making model was applied. This approach allows policy makers to integrate worldviews, culture, diverse flexibility of concerned communities and other stakeholders, perceptions, values, attitudes and behaviors in policy making.

Author Contributions: V.E.N.: conceptualization, methodology, data curation, writing—original draft preparation. C.N.M.: supervision, validation, writing—review and editing. I.C.E.: project administration, resources, visualization. All authors have read and agreed to the published version of the manuscript.

Funding: This research received no external funding.

Institutional Review Board Statement: Not applicable.

Informed Consent Statement: Not applicable.

Conflicts of Interest: The authors declare that they have no known competing financial interests or personal relationships that could have appeared to influence the work reported in this paper.

Appendix A

Appendix A.1 Geometric Means and Priority Indices Section A

Table A1. Geometric means and the priority indices of the Biodiversity Conservation Approa hes (BCAs) both in expectations and current performance.

a: Geometric mean of the three BCAs on expectations/capacity building			
	Species-based (BCA1)	Area-based (BCA2)	Ecosystem-service-based (BCA3)
Species-based (BCA1)	1	0.42	0.65
Area-based (BCA2)	2.40	1	0.57
Ecosystem-service-based (BCA3)	1.53	1.76	1

b: Row average operation of the three BCAs on expectations/capacity building			
	Species-based (BCA1)	Area-based (BCA2)	Ecosystem-service-based (BCA3)	Priority Index (Row Average)
Species-based (BCA1)	0.203	0.132	0.293	0.209
Area-based (BCA2)	0.487	0.314	0.257	0.353
Ecosystem-service-based (BCA3)	0.310	0.553	0.450	0.438

c: Geometric mean of the three BCAs on actual/current performance			
	Species-based (BCA1)	Area-based (BCA2)	Ecosystem-service-based (BCA3)
Species-based (BCA1)	1	0.54	0.85
Area-based (BCA2)	1.88	1	0.37
Ecosystem-service-based (BCA3)	1.18	2.72	1

d: Row average operation of the three BCAs on actual/current performance			
	Species-based (BCA1)	Area-based (BCA2)	Ecosystem-service-based (BCA3)	Priority Index (Row Average)
Species-based (BCA1)	0.246	0.127	0.383	0.252
Area-based (BCA2)	0.463	0.235	0.167	0.288
Ecosystem-service-based (BCA3)	0.291	0.638	0.451	0.460

Appendix A.2 Geometric Means and Priority Indices Section B

Table A2. Geometric means and the priority indices of the Conservation Targets (CTs) both in expectations and current performance.

a: Geometric mean of the four CTs of the species-based approach on expectations/capacity building				
	Threatened species (CT1)	Ecological important species (hubs of network) (CT2)	Species of use to human (CT3)	Species with non-use values (CT4)
Threatened species (CT1)	1	0.59	0.36	1.79
Ecological important species (hubs of network) (CT2)	1.70	1	0.32	0.97
Species of use to human (CT3)	2.77	3.17	1	1.61
Species with non-use values (CT4)	0.56	1.03	0.62	1

b: Row average operation of the four CTs of the species-based approach on expectations/capacity building					
	Threatened species (CT1)	Ecological important species (hubs of network) (CT2)	Species of use to human (CT3)	Species with non-use values (CT4)	Priority Index (Row Average)
Threatened species (CT1)	0.166	0.102	0.157	0.333	0.190
Ecological important species (hubs of network) (CT2)	0.282	0.173	0.139	0.181	0.194
Species of use to human (CT3)	0.459	0.547	0.435	0.300	0.435
Species with non-use values (CT4)	0.093	0.178	0.270	0.186	0.82

Table A2. *Cont.*

c: Geometric mean of the four CTs of the species-based approach on actual/current performance				
	Threatened species (CT1)	Ecological important species (hubs of network) (CT2)	Species of use to human (CT3)	Species with non-use values (CT4)
Threatened species (CT1)	1	0.63	0.49	1.28
Ecological important species (hubs of network) (CT2)	1.59	1	0.32	1.12
Species of use to human (CT3)	2.04	3.13	1	1.22
Species with non-use values CT4	0.78	0.89	0.82	1

d: Row average operation of the four CTs of the species-based approach on actual/current performance					
	Threatened species (CT1)	Ecological important species (hubs of network) (CT2)	Species of use to human (CT3)	Species with non-use values (CT4)	Priority Index (Row Average)
Threatened species (CT1)	0.185	0.112	0.186	0.277	0.190
Ecological important species (hubs of network) (CT2)	0.294	0.177	0.122	0.242	0.209
Species of use to human (CT3)	0.377	0.550	0.380	0.264	0.393
Species with non-use values (CT4)	0.014	0.158	0.312	0.216	0.175

e: Geometric mean of the two CTs of the area-based approach on expectation/capacity building		
	Endemic areas (hotspots) CT5)	Non-endemic areas (CT6)
Endemic areas (hotspots) (CT5)	1	1.54
Non-endemic areas (CT6)	0.65	1

f: Row average operation of the two CTs of the area-based approach on expectation/capacity building			
	Endemic areas (hotspots) (CT5)	Non-endemic areas (CT6)	Priority Index (Row Average)
Endemic areas (hotspots) (CT5)	0.606	0.606	0.606
Non-endemic areas (CT6)	0.394	0.394	0.394

g: Geometric mean of the two CTs of the area-based approach on actual/current performance		
	Endemic areas (hotspots) (CT5)	Non-endemic areas (CT6)
Endemic areas (hotspots) (CT5)	1	1.00
Non-endemic areas (CT6)	1.00	1

h: Row average operation of the two CTs of the area-based approach on actual/current performance			
	Endemic areas (hotspots) (CT5)	Non-endemic areas (CT6)	Priority Index (Row Average)
Endemic areas (hotspots) (CT5)	0.5	0.5	0.5
Non-endemic areas (CT6)	0.5	0.5	0.5

i: Geometric mean of the three CTs of the ecosystem-service-based approach on expectations/capacity building			
	Water (CT7)	Land (CT8)	Living Resources (CT9)
Water (CT7)	1	0.28	0.43
Land (CT8)	3.59	1	0.47
Living Resources (CT9)	2.32	2.15	1

j: Row average operation of the three CTs of the ecosystem-service-based approach on expectation/capacity building				
	Water (CT7)	Land (CT8)	Living Resources (CT9)	Priority Index (Row Average)
Water (CT7)	0.104	0.154	0.104	0.204
Land (CT8)	0.374	0.194	0.114	0.251
Living Resources (CT9)	0.242	0.417	0.243	0.273

Table A2. *Cont.*

k: Geometric mean of the three CTs of the ecosystem-service-based approach on actual/current performance			
	Water (CT7)	Land (CT8)	Living Resources (CT9)
Water (CT7)	1	1.19	0.84
Land (CT8)	0.84	1	0.85
Living Resources (CT9)	1.19	1.18	1
Column Total	3.03	3.37	2.69

l: Row average operation of the three CTs of the ecosystem-service-based approach on actual/current performance			
	Water (CT7)	Land (CT8)	Living Resources (CT9)	Priority Index (Row Average)
Water (CT7)	0.213	0.288	0.230	0.121
Land (CT8)	0.255	0.237	0.262	0.247
Living Resource (CT9)	0.253	0.286	0.273	0.268

Appendix B

Step 11: Establish Input–Output Relationship of the Biodiversity Conservation Approaches

Practically, the importance of this model is utilized here to show such interdependence using:

$$x = (I - A)^{-1}d \tag{A1}$$

where

x = vector of total output (dependence vector);
I = identity matrix;
A = matrix of coefficients a_{ij} (geometric mean of the Delphi data);
d = vector of final demand (α matrix which is the priority indices on expectations of the BCAs).

Since the interflow matrices $(I - A)$ have a multiplicative inverse, then this depicts a linear system of equations having a unique solution, and so, given final demand vectors, we find the needed output.

Appendix C

Step 12: Formulating Linear Programming (LP)

Let r_k represent the resource need for BCAk where k (the decision variables) = 1, 2, 3.

$$R = \sum_{k=1}^{3} r_k \tag{A2}$$

where R = objective function and r = coefficient of objective function corresponding to the decision variables.

Then:

$$w_k = \frac{r_k}{R} \tag{A3}$$

where w_k = the vector of the decision variables to be determined.

The specific objective here is to maximize the use of resources for the BCA in order to minimize depletion of biological diversity. Therefore, following Satty and Alexander (1981), Madu and Madu (1993) and Madu et al. (2017), an LP model can be established more compactly as:

$$\text{Max } \beta^{\mathrm{T}} w \tag{A4}$$

Subject to:

$$0 \leq w_k \leq \frac{r_k}{R} \tag{A5}$$

and

$$\sum_{k=1}^{3} w_k = 1 \tag{A6}$$

where w is a real matrix that corresponds to the coefficients on k_1, k_2 and k_3 in the constrains of the LP problem, and Max $\beta^T w$ is the objective function.

References

1. Josh, D.; Ian, O.; Julian, W. Consensus when experts disagree: A priority list of invasive alien plant species that reduce ecological restoration success. *Manag. Biol. Invasions* **2018**, *9*, 329–341.
2. Spies, T.A.; Ripple, W.J.; Bradshaw, G.A. Dynamics and pattern of a managed coniferous forest landscape in Oregon. *Ecol. Appl.* **1994**, *4*, 555–568. [CrossRef]
3. Lathrop, R.G.; Bognar, J.A. Applying GIS and landscape ecological principles to evaluate land conservation alternatives. *Landsc. Urban Plan.* **1998**, *41*, 27–41. [CrossRef]
4. Bradley, J.C.; Duffy, J.E.; Gonzalez, A.; Hooper, D.U.; Perrings, C.; Venail, P.; Narwani, A.; Mace, G.M.; Tilman, D.; Wardle, D.A.; et al. Biodiversity Loss and Its Impact on Humanity. *Nature* **2012**, *486*, 59–67. [CrossRef]
5. Sam, I.E.; Nnaji, E.S.; Asuquo, E.E.; Etefia, T.E. Impact of Overpopulation on the Biological Diversity Conservation in Boki Local Government Area of Cross River State. *Niger. Am. J. Environ. Eng.* **2014**, *4*, 94–98. Available online: http://article.sapub.org/10.5923.j.ajee.20140405.02.html (accessed on 26 October 2018).
6. Regan, H.; Davis, F.; Andelman, S.; Widyanata, A.; Freese, M. Comprehensive criteria for biodiversity evaluation in conservation planning. *Biodivers. Conserv.* **2007**, *16*, 2715–2728. [CrossRef]
7. Pearce, D.; Moran, D. *The Economic Value of Biodiversity*; Earthscan: London, UK, 1994; p. 186.
8. Costanza, R.; D'arge, R.; De Groot, R.; Farber, S.; Grasso, M.; Hannon, B.; Limburg, K.; Naeem, S.; O'neil, R.V.; Paruelo, J.; et al. The value of the world's ecosystem services and natural capital. *Nature* **1997**, *387*, 253–260. [CrossRef]
9. Jones, P.G.; Beebe, S.E.; Tohme, J.; Galwey, N.W. The use of geographic information systems in biodiversity exploration and conservation. *Biodivers. Conserv.* **1997**, *6*, 947–958. [CrossRef]
10. Mariuki, J.N.; De Klerk, H.M.; Williams, P.H.; Bennun, L.A.; Crowe, T.M.; Berge, E.V. Using patterns of distribution and diversity of Kenyan birds to select and prioritize areas for conservation. *Biodivers. Conserv.* **1997**, *6*, 191–210. [CrossRef]
11. Rubino, M.J.; Hess, G.R. Planning open spaces for wildlife 2: Modeling and verifying focal species habitat. *Landsc. Urban Plan.* **2003**, *64*, 89–104. [CrossRef]
12. Smallwood, K.S.; Wilcox, B.; Leidy, R.; Yarris, K. Indicators assessment for habitat conservation plan of Yolo County, California, USA. *Environ. Manag.* **1998**, *22*, 947–958. [CrossRef]
13. Roberts, C.M.; Andelman, S.; Branch, G.; Bustamante, R.H.; Carlos Castilla, J.; Dugan, J.; Halpern, B.S.; Lafferty, K.D.; Leslie, H.; Lubchenco, J. Ecological criteria for evaluating candidate sites for marine reserves. *Ecol. Appl.* **2003**, *13*, 199–214. [CrossRef]
14. Gilman, E.; Dunn, D.; Read, A.; Hyrenbach, K.D.; Warner, R. Designing criteria suites to identify discrete and networked sites of high value across manifestations of biodiversity. *Biodivers. Conserv.* **2011**, *20*, 3363–3383. [CrossRef]
15. Bidegain, I.; Cerda, C.; Catala'n, E.; Tironi, A.; Lo'pez-Santiago, C. Social preferences for ecosystem services in a biodiversity hotspot in South America. *PLoS ONE* **2019**, *14*, e0215715. [CrossRef]
16. Jacobs, S.; Dendoncker, N.; Martı'n-Lo'pez, B.; Barton, D.N.; Gomez-Baggethun, E.; Boeraeve, F.; McGrath, F.L.; Vierikko, K.; Geneletti, D.; Sevecke, K.J.; et al. A new valuation school: Integrating diverse values of nature in resource and land use decisions. *Ecosyst. Serv.* **2016**, *22*, 213–220. [CrossRef]
17. López-Santiago, C.A.; Oteros-Rozas, E.; Martin-López, B.; Plieninger, T.; Martin, E.G.; González, J.A. Using visual stimuli to explore the social perceptions of ecosystem services in cultural landscapes: The case of transhumance in Mediterranean Spain. *Ecol. Soc.* **2014**, *19*, 27. [CrossRef]
18. Ode, A.; Fry, G.; Tveit, M.S.; Messager, P.; Miller, D. Indicators of perceived naturalness as drivers of landscape preference. *J. Environ. Manag.* **2009**, *90*, 375–383. [CrossRef]
19. Nijnik, M.; Zahvoyska, L.; Nijnik, A.; Ode, A. Public evaluation of landscape content and change: Several examples from Europe. *Land Use Policy* **2008**, *26*, 77–86. [CrossRef]
20. Tveit, M.; Ode, A.; Fry, G. Key concepts in a framework for analysing visual landscape character. *Landsc. Res.* **2006**, *31*, 229–255. [CrossRef]
21. Martín-López, B.; Iniesta-Arandia, I.; García-Llorente, M.; Palomo, I.; Casado-Arzuaga, I.; Amo, D.G.; Gómez-Baggethun, E.; Oteros-Rozas, E.; Palacios-Agundez, I.; Willaarts, B.; et al. Uncovering ecosystem service bundles through social preferences. *PLoS ONE* **2012**, *7*, e38970. [CrossRef]
22. Vane-Wright, R.I.; Humphries, C.J.; Williams, P.H. What to protect—systematics and the agony of choice. *Biol. Conserv.* **1991**, *55*, 235–254. [CrossRef]
23. Roberge, J.M.; Angelstam, P. Usefulness of the umbrella species concept as a conservation tool. *Conserv. Biol.* **2004**, *18*, 76–85. [CrossRef]
24. de Lima, R.A.F.; Souza, V.C.; de Siqueira, M.F.; ter Steege, H. Defining endemism levels for biodiversity conservation: Tree species in the Atlantic Forest hotspot. *Biol. Conserv.* **2020**, *252*, 108825. [CrossRef]

25. Moreira, F.; Allsopp, N.; Esler, K.J.; Wardell-Johnson, G.; Ancillotto, L.; Arianoutsou, M.; Clary, J.; Brotons, L.; Clavero, M.; Dimitrakopoulos, P.G.; et al. Priority questions for biodiversity conservation in the Mediterranean biome: Heterogeneous perspectives across continents and stakeholders. *Conserv. Sci. Pract.* **2019**, *1*, e118. [CrossRef]
26. Bosso, L.; Ancillotto, L.; Smeraldo, S.; D'Arco, S.; Migliozzi, A.; Conti, P.; Russo, D. Loss of potential bat habitat following a severe wildfire: A model based rapid assessment. *Int. J. Wildland Fire* **2018**, *27*, 756–769. [CrossRef]
27. Smeraldo, S.; Bosso, L.; Fraissinet, M.; Bordignon, L.; Brunelli, M.; Ancillotto, L.; Russo, D. Modelling risks posed by wind turbines and electric power lines to soaring birds: The black stork (Ciconia nigra) in Italy as a case study. *Biodivers. Conserv.* **2020**, *29*, 1959–1976. [CrossRef]
28. Chandra, N.; Singh, G.; Lingwal, S.; Jalal, J.S.; Bisht, M.S.; Pal, V.; Rawat, B.; Tiwari, L. Ecological niche modeling and status of threatened alpine medicinal plant Dactylorhiza Hatagirea, D. Don in Western Himalaya. *J. Sustain. For.* **2021**, 1–17. [CrossRef]
29. Shmelev, S.E. Working Paper Number 181. In *Multi-criteria Assessment of Ecosystems and Biodiversity: New Dimensions and Stakeholders in the South of France*; University of Oxford: Oxford, UK, 2010.
30. Madu, C.N.; Madu, A.A. A systems approach to the transfer of mutually dependent Technologies. *Socio-Econ. Plan. Sci.* **1993**, *27*, 269–287. [CrossRef]
31. Madu, C.N.; Ozumba, B.C.; Kuei, C.-H.; Madu, I.E.; Nnadi, V.E.; Ezeasor, I.C.; Eze, H.I.; Odinkonigbo, U.; Ogbuene, E.; Ewoh, J.; et al. Hyogo Framework for Action (HFA): A case study of Nigeria. *Curr. Environ. Manag.* **2019**, *6*, 196–209. [CrossRef]
32. Madu, C.N.; Kuei, C.H.; Madu, I.E.; Ozumba, B.C.; Odinkonigbo, I.L.; Ezeasor, I.C.; Nnadi, V.E.; Ogbuene, E.B.; Ewoh, J.C.; Mmomh, P.C. Priority Assessment of the Hyogo Framework for Action (HFA): A case study of Nigeria. In *Hand Book of Disaster Risk Reduction and Management*, 1st ed.; Madu, C.N., Kuei, C.H., Eds.; World Scientific Publishing: Hackensack, NJ, USA, 2017.
33. Schmidt, K.; Aumann, I.; Hollander, I.; Damm, K.; Schulenburg, J.G. Applying the Analytic Hierarchy Process in healthcare research: A systematic literature review and evaluation of reporting. *BMC Med. Inform. Decis. Mak.* **2015**, *15*, 112. [CrossRef]
34. Joshi, V.; Narra, V.R.; Joshi, K.; Lee, K.; Melson, D. PACS Administrators' and Radiologists' Perspective on the Importance of Features for PACS Selection. *J. Digit. Imaging* **2014**, *27*, 486–495. [CrossRef]
35. Cheever, M.A.; Allison, J.P.; Ferris, A.S.; Finn, O.J.; Hastings, B.M.; Hecht, T.T.; Mellman, I.; Prindiville, S.A.; Viner, J.L.; Weiner, L.M.; et al. The prioritization of cancer antigens: A national cancer institute pilot project for the acceleration of translational research. *Clin. Cancer Res.* **2009**, *15*, 5323–5337. [CrossRef]
36. Liberatore, M.J.; Nydick, R.L. The analytic hierarchy process in medical and health care decision making: A literature review. *Eur. J. Oper. Res.* **2008**, *189*, 194–207. [CrossRef]
37. Kleindorfer, P.R.; Partovi, F.Y. Integrating Manufacturing strategy and technology choice. *Eur. J. Oper. Res.* **1990**, *47*, 139–270. [CrossRef]
38. Marcot, B.G.; Thompson, M.P.; Runge, M.C.; Thompson, F.R.; Cleaves, D.; Tomosy, M.; Fisher, L.A.; Bliss, A. Recent advances in applying decision science to managing national forests. *For. Ecol. Manag.* **2012**, *285*, 123–132. [CrossRef]
39. Asaad, I.; Lundquist, C.J.; Erdmann, M.V.; Costello, M.J. Ecological criteria to identify areas for biodiversity conservation. *Biol. Conserv.* **2017**, *213*, 309–316. [CrossRef]
40. Pereira, H.M.; Navarro, L.M.; Martins, I.S. Global biodiversity change: The bad, the good, and the unknown. *Annu. Rev. Environ. Resour.* **2012**, *37*, 25–50. [CrossRef]
41. Tongco, M.D.C. Purposive Sampling as a Tool for Informant Selection. *Ethnobot. Res. Appl.* **2007**, *5*, 147–158. Available online: https://scholarspace.manoa.hawaii.edu/bitstream/10125/227/4/I1547-3465-05-147.pdf (accessed on 15 August 2021). [CrossRef]
42. Powell, C. The Delphi technique: Myths and realities. *J. Adv. Nurs.* **2003**, *41*, 376–382. [CrossRef]
43. Besaw, L.E.; Rizzo, D.M.; Kline, M.; Underwood, K.L.; Doris, J.J.; Morrissey, L.A.; Pelletier, K. Stream classification using hierarchical artificial neural networks: A fluvial hazard management tool. *J. Hydrol.* **2009**, *373*, 34–43. [CrossRef]
44. Saffron, J.O.; Tim, J.O.; Mike, H.; Irene, L.; Andrew, R.W. Using expert knowledge to assess uncertainties in future polar bear populations under climate change. *J. Appl. Ecol.* **2008**, *45*, 1649–1659. Available online: https://besjournals.onlinelibrary.wiley.com/doi/pdf/10.1111/j.1365-2664.2008.01552.x (accessed on 15 August 2021).
45. Wooldridge, S.; Done, T.; Berkelmans, R.; Jones, R.; Marshall, P. Precursors for resilience in coral communities in a warming climate: A belief network approach. *Mar. Ecol. Prog. Ser.* **2005**, *295*, 157–169. [CrossRef]
46. Krayer von Krauss, M.P.; Casman, E.A.; Small, M.J. Elicitation of expert judgments of uncertainty in the risk assessment of herbicide-tolerant oilseed crops. *Risk Anal.* **2004**, *24*, 1515–1527. [CrossRef] [PubMed]
47. Mariëlle, Z.; Wenneke, S.H.; Herbert, H.; Rens, S. Application and Evaluation of an Expert Judgement Elicitation Procedure for Correlations. *Front. Psychol.* **2017**, *8*, 90. Available online: https://journals.plos.org/plosone/article/file?id=10.1371/journal.pone.0198468&type=printable (accessed on 15 August 2021).
48. Murray, J.V.; Goldizen, A.W.; O'Leary, R.A.; McAlpine, C.A.; Possingham, H.P.; Low Choy, S. How useful is expert opinion for predicting the distribution of a species within and beyond the region of expertise? A case study using brushtailed rock-wallabies Petrogalepenicillata. *J. Appl. Ecol.* **2009**, *46*, 842–851. [CrossRef]
49. Martin, T.G.; Kuhnert, P.M.; Mengersen, K.; Possingham, H.P. The power of expert opinion in ecological models using Bayesian methods: Impact of grazing on birds. *Ecol. Appl.* **2005**, *15*, 266–280. [CrossRef]
50. Brouwer, R.; De Blois, C. Integrated modelling of risk and uncertainty underlying the cost and effectiveness of water quality measures. *Environ. Model. Softw.* **2008**, *23*, 922–937. [CrossRef]

51. Fish, R.; Winter, M.; Oliver, D.M.; Chadwick, D.; Selfa, T.; Heathwaite, A.L.; Hodgson, C. Unruly pathogens: Eliciting values for environmental risk in the context of heterogeneous expert knowledge. *Environ. Sci. Policy* **2009**, *12*, 281–296. [CrossRef]
52. Montangeroa, A.; Belevib, H. Assessing nutrient flows in septic tanks by eliciting expert judgement: A promising method in the context of developing countries. *Water Res.* **2007**, *41*, 1052–1064. [CrossRef]
53. Aczel, J.; Saaty, T. Procedures for Synthesizing Ratio Judgments. *J. Math. Psychol.* **1983**, *27*, 93–102. [CrossRef]
54. Brans, J.P.; Vincke, P.; Mareschal, B. How to select and how to rank projects: The promethee method. *Eur. J. Oper. Res.* **1986**, *24*, 228–238. [CrossRef]
55. Rowe, G.; Wright, G. Expert Opinions in Forecasting. Role of the Delphi Technique. In *Principles of Forecasting: A Handbook of Researchers and Practitioners*; Armstrong, J.S., Ed.; Kluwer Academic Publishers: Boston, MA, USA, 2001.
56. Fink, A.; Kosecoff, J.; Chassin, M.; Brook, R. *Consensus Methods: Characteristics and Guidelines for Use*; RAND: Santa Monica, CA, USA, 1991.
57. Hsu, C.; Sandford, B.A. The Delphi Technique: Making Sense of Consensus. *Pract. Assess. Res. Eval.* **2007**, *12*, 10. [CrossRef]
58. MEA. *Ecosystems and Human Well-Being: Synthesis Report*; Island Press: Washington, DC, USA, 2005.
59. Seppelt, R.; Dormann, C.F.; Eppink, F.V.; Lautenbach, S.; Schmidt, S. A quantitative review of ecosystem service studies: Approaches, shortcomings and the road ahead. *J. Appl. Ecol.* **2011**, *48*, 630–636. Available online: http://onlinelibrary.wiley.com/doi/10.1111/j.1365-2664.2010.01952.x/full (accessed on 15 August 2021). [CrossRef]
60. Franklin, J.F. Preserving biodiversity: Species, ecosystems, or landscapes? *Ecol. Appl.* **1993**, *3*, 202–205. [CrossRef]
61. Vuilleumier, S.; Prelaz-Droux, R. Map of ecological networks for landscape planning. *Landsc. Urban Plan.* **2002**, *58*, 157–170. [CrossRef]
62. Phua, M.; Minowa, M.A. GIS-based multi-criteria decision making approach to forest conservation planning at a landscape scale: A case study in the Kinabalu Area, Sabah, Malaysia. *Landsc. Urban Plan.* **2005**, *71*, 207–222. Available online: www.elsevier.com/locate/landurbplan (accessed on 16 August 2017).
63. Washington Biodiversity Council. Washington Biodiversity Conservation Strategy: Sustaining Our Natural Heritage for Future Generations. 2007. Available online: http://www.rco.wa.gov/documents/biodiversity/WABiodiversityConservationStrategy.pdf (accessed on 5 September 2017).
64. UNCED. The United Nations Conference on Environment and Development. 1992. Available online: http://www.unesco.org/education/pdf/RIO_E.PDF (accessed on 16 August 2017).
65. MEA. *Ecosystems and Human Well-Being—A Framework for Assessment*; World Resources Institute Island Press: New York, NY, USA, 2003.
66. Myers, N.; Mittermeier, R.A.; Mittermeier, C.G.; da Fonseca, G.A.B.; Kent, J. Biodiversity hotspots for conservation priorities. *Nature* **2000**, *403*, 853–858. [CrossRef]
67. Barends, E.; Rousseau, D.M.; Briner, R.B. Evidence-Based Management: The Basic Principles. Amsterdam: Center for Evidence-Based Management. 2014. Available online: https://www.cebma.org/wpcontent/uploads/Evidence-Based-Practice-The-Basic-Principles.pdf (accessed on 16 August 2017).
68. Saaty, T.L.; Alexander, J.M. *Thinking with Models*; Pergamon Press: Oxford, UK, 1981.
69. Schrade, L. *User's Manual for the Linear, integer, and Quadratic Programming with LINDO*; Scientific Press: Redwood City, CA, USA, 1987.

 *sustainability*

Article

Multi-Disciplinary Assessment of Napier Grass Plantation on Local Energetic, Environmental and Socioeconomic Industries: A Watershed-Scale Study in Southern Thailand

Kotchakarn Nantasaksiri [1],*, Patcharawat Charoen-amornkitt [2], Takashi Machimura [1] and Kiichiro Hayashi [3]

1   Division of Sustainable Energy and Environmental Engineering, Graduate School of Engineering, Osaka University, Osaka 565-0871, Japan; mach@see.eng.osaka-u.ac.jp

2   Department of Mechanical Engineering, Faculty of Engineering, King Mongkut's University of Technology Thonburi, Bangkok 10140, Thailand; patcharawat.c@gmail.com

3   Institute of Materials and Systems for Sustainability, Nagoya University, Nagoya 464-8601, Japan; maruhaya98–@nagoya-u.jp

*   Correspondence: kotchakarn@ge.see.eng.osaka-u.ac.jp

Abstract: Napier grass is an energy crop that is promising for future power generation. Since Napier grass has never been planted extensively, it is important to understand the impacts of Napier grass plantations on local energetic, environmental, and socioeconomic features. In this study, the soil and water assessment tool (SWAT) model was employed to investigate the impacts of Napier grass plantation on runoff, sediment, and nitrate loads in Songkhla Lake Basin (SLB), southern Thailand. Historical data, collected between 2009 and 2018 from the U-tapao gaging station located in SLB were used to calibrate and validate the model in terms of precipitation, streamflow, and sediment. The simulated precipitation, streamflow, and sediment showed agreement with observed data, with the coefficients of determination being 0.791, 0.900, and 0.997, respectively. Subsequently, the SWAT model was applied to evaluate the impact of land use change from the baseline case to Napier grass plantation cases in abandoned areas with four different nitrogen fertilizer application levels. The results revealed that planting Napier grass decreased the average surface runoff and sediment in the watershed. A multidisciplinary assessment supporting future decision making was conducted using the results obtained from the SWAT model; these showed that Napier grass will provide enhanced benefits to hydrology and water quality when nitrogen fertilizers of 0 and 125 kgN ha^{-1} were applied. On the other hand, the benefits to the energy supply, farmer's income, and CO_2 reduction were highest when a nitrogen fertilization of 500 kgN ha^{-1} was applied. Nonetheless, planting Napier grass should be supported since it increases the energy supply and creates jobs while also reducing surface runoff, sediment yield, nitrate load, and CO_2 emission.

Keywords: SWAT model; water quality; hydrology; fertilizer application; Songkhla Lake Basin

Citation: Nantasaksiri, K.; Charoen-amornkitt, P.; Machimura, T.; Hayashi, K. Multi-Disciplinary Assessment of Napier Grass Plantation on Local Energetic, Environmental and Socioeconomic Industries: A Watershed-Scale Study in Southern Thailand. *Sustainability* **2021**, *13*, 13520. https://doi.org/10.3390/su132413520

Academic Editor: Christian N. Madu

Received: 5 November 2021
Accepted: 2 December 2021
Published: 7 December 2021

Publisher's Note: MDPI stays neutral with regard to jurisdictional claims in published maps and institutional affiliations.

1. Introduction

The increase in fossil-fuel-based energy consumption has resulted in an energy crisis and a critical state of global warming in the 21st century. Finding a new energy source that is renewable, clean, and sustainable has thus become an urgent task. Among the candidates suitable for such clean energy, biogas, which is a gaseous fuel obtained from the decomposition of organic matter, has received widespread attention [1–3]. Biogas feedstocks can be found in a variety of forms, such as agricultural residues, forestry byproducts, animal waste, and dedicated energy crops. Biogas energy derived from dedicated energy crops can be considered carbon-neutral because although carbon is released during the process of power generation, the same amount of carbon is absorbed by the crops during their growth. The advantages of biogas are not limited to carbon neutrality—it also has higher yields and a shorter life cycle, thus promising a stable fuel supply. However, for the widespread use of these renewable energy crops, there are several aspects to be considered prior to making decisions regarding land use change for energy plantations, in order to avoid unforeseen socioeconomic and environmental issues [4–7].

Utilizing bioenergy can directly satisfy the Sustainable Development Goals (SDGs) suggested by the United Nations Development Programme (UNDP), namely, SDG 7 (affordable and clean energy), SDG 8 (decent work and economic growth), and SDG 13 (climate action). However, there are several SDGs that are indirectly related to the use of bioenergy, since planting new dedicated energy crops would affect water and land. For example, SDG 6 ensures safe and affordable drinking water and freshwater supplies, while SDG 15 conserves and restores terrestrial and freshwater ecosystems. This highlights the importance of a multidisciplinary assessment for the safe introduction of bioenergy as an alternative energy source.

Napier grass (*Cenchrus purpureus Schumach.*) is one of the popular perennial grasses used in biogas-based power generation [8–13]. It possesses many desirable characteristics for energy crops, such as a short life cycle, relatively high methane content, and high water use efficiency [14–16]. Additionally, it has been reported that perennial grasses can help reduce nitrate transport into the soil, which is a waterborne pollutant [17]. Furthermore, Napier grass can grow well under flooded soil conditions, making it suitable for use in water pollution treatment [18]. Although Napier grass can serve as an alternative fuel source and reduce carbon emissions, there are concerns over the impacts of such a crop on the local soil and water, due to changes in land use and intensive agricultural practices [19–21].

To investigate the impacts of land use changes on the local soil and water, the soil and water assessment tool (SWAT) model is one of the most promising models. Dos Santos et al. [22] utilized the SWAT model to investigate the impacts of land use changes on streamflow and sediment yield and discuss ways to consider future land use conditions in the Atibiai River basin, Brazil. The results provided useful information for proposing improvements in the basin's environmental quality and management. The SWAT model was also applied to assess the impact of changes in agricultural management practices on nitrate loads by Epelde et al. [23]. They found that the trends of nitrogen surplus in the system generally increased as the fertilization input increased. The effects of replacing conventional crops with Miscanthus on riverine nitrate load were investigated by Ng et al. [24] using the SWAT model. The results revealed that the nitrate load tended to decrease when replacing conventional crops with Miscanthus. Similarly, using the SWAT model, Cibin et al. [25] investigated the impacts of bioenergy crops on hydrology and water quality. The study also found that perennial grass reduced pollutant load at the watershed outlet. This suggested that the study on the impacts of land use changes on the local soil and water is currently of interest.

In our previous study that evaluated the land potential for Napier grass cultivation, the dry matter yield (DMY) was estimated using the SWAT model [26]. The SWAT model successfully estimated the Napier grass DMY in Thailand, with a coefficient of determination of 0.951; the results also show that southern Thailand had the highest average DMY. In our continued study [27], we integrated the land suitability map, obtained from a multicriteria decision analysis, and the spatial distribution of DMY to assess the suitability of a site for Napier-grass-based biogas power plants in southern Thailand. The location of biogas power plants and their distance from roads, residential areas, waterbodies, their access to the electricity grid, and their supply of feedstock were all considered during the site suitability analysis. The results revealed that, using only Napier grass from abandoned areas, five biogas power plants could be built with a total contracted capacity of 420 MW. This highlights that Napier grass can significantly reduce Thailand's dependency on imported electricity.

Although Nantasaksiri et al. [26,27] suggested that Napier grass possesses massive potential in biogas-based power generation in southern Thailand, with a few socioeconomic and environmental criteria considered, a study on the impacts of Napier grass on hydrology and water quality is yet to be performed. Moreover, the factors to be considered should not be limited to hydrology and water quality because there are several varied parties involved: The government will likely focus on energy supply, CO_2 reduction, and sustainability, while the local community and farmers are likely most interested in job creation and income. Hence, the broad objective of this study was to investigate the effects of different management practices for Napier grass plantations on surface runoff, sediment yield, and nitrate load in southern Thailand using the SWAT model. Based on the results obtained, a multidisciplinary assessment for supporting adequate decision making to utilize Napier grass as a feedstock for biogas-based power generation was carried out to comparatively assess the advantages and disadvantages of various cases of cultivation practices with different fertilizing levels. The results from this study provide a logical framework to support decision making for implementing new dedicated energy crops as a biogas feedstock, which is useful for the transition toward a renewable and sustainable energy society.

2. Materials and Methods

2.1. Study Site

The Songkhla Lake Basin (SLB), shown in Figure 1, was selected as the study site since our previous study had found three potential sites suitable for biogas-based power plants here [27]. This basin is located in southern Thailand and lies within three provinces, namely, Phattalung, Songkhla, and Nakhonsithammarat, and has an area of approximately 8157 km². The elevation of the watershed ranges from 0 to 1334 m above sea level, and the average annual precipitation is 1992 mm. In this watershed, the annual average temperature is 27.4 °C; the highest average temperature (33.9 °C) was observed in April, and in October, the average temperature was found to be lowest (23.0 °C). The average relative humidity for the SLB was 81.0%. The major land uses in the basin are agricultural (60.46%), forest (13.79%), and water bodies (13.54%). Southern Thailand is famous for latex production, and 41.33% of the total area in this basin is utilized for the Pará rubber tree plantation. The largest natural lake in Thailand, i.e., Songkhla Lake, is located in the SLB. Among the chain of lagoons that form Songkhla Lake, the northernmost lagoon, i.e., Thale Noi, was declared a protected freshwater wetland in 1975. In addition, the Kuan Ki Sian knoll in the non-hunting area was declared a wetland of international importance in 1998 by the Ramsar Convention. Approximately 1.5 million people live in the basin, resulting in the rapid degradation of natural resources in the area because of economic activities. Because many parts of the SLB are wetlands, the area is highly susceptible to flooding and landslides, and several studies have focused on combating these issues [28,29].

2.2. Model Description

The SWAT model, a continuous time- and process-based watershed model, was used to examine the impacts of Napier grass plantations on the hydrology and water quality within SLB. A detailed description of the SWAT model used for calculating Napier grass DMY can be found in our previous study [26]. Based on the SWAT theoretical documentation [30], surface runoff, sediment yield, and nitrate load, which are considered indicators of environmental burdens corresponding to flood, erosion, and water pollution, respectively, were calculated. The surface runoff Q_{surf} (mm day^{-1}) is described using the following equation:

$$Q_{surf} = \frac{\left(R_{day} - 0.2S\right)^2}{R_{day} + 0.8S} \tag{1}$$

where R_{day} is the rainfall depth for the day (mm day^{-1}) and S is the retention parameter (mm day^{-1}), which is defined as:

$$S = 25.4\left(\frac{1000}{CN} - 10\right) \tag{2}$$

where CN is the curve number for the day. The curve number CN is an important parameter for calculating surface runoff, depending on the hydrologic soil group and land use. In the SWAT model, a higher curve number indicates a higher runoff potential, and a lower number indicates greater retention.

To calculate sediment yield *sed* (in metric tons), the modified Universal Soil Loss Equation (MUSLE) equation was used. In this approach, the sediment yield is a function of the surface runoff Q_{surf}. The MUSLE equation can be expressed as:

$$sed = 11.8 \left(Q_{surf} q_{peak} A_{hru} \right)^{0.56} K \cdot C \cdot P \cdot LS \cdot CFRG \tag{3}$$

where q_{peak} is the peak runoff rate (m^3 s^{-1}), A_{hru} is the area of the hydrological response unit (HRU) (ha), K is the USLE soil erodibility factor, C is the USLE cover and management factor, P is the USLE support practice factor, LS is the USLE topographic factor, and $CFRG$ is the coarse fragment factor.

The total nitrate content assessed in the SWAT model is an integrated contribution of fertilizer, manure application, bacterial attachment, mineralization, atmospheric deposition, plant uptake, leaching, volatilization, denitrification, and erosion. Because there are several equations involving calculations of the nitrate cycle, the SWAT model's description was carefully summarized by Hass et al. [31], which will not be repeated here. The total nitrate balance in a period, ΔN (kgN ha^{-1}), was calculated using Equation (4), i.e., the difference between nitrate input into the nitrate pool and the nitrate used by agricultural activities:

$$N = \left(N_{fert} + N_{hum} + N_{min} + N_{atm} \right) - \left(N_{denit} + N_{up} + N_{leach} + N_{surf} + N_{latf} \right) \tag{4}$$

where N_{fert} is the amount of nitrate in fertilizers, N_{hum} is the nitrogen mineralization from the humus active organic nitrogen pool (the amount of nitrogen moving from the active organic to nitrate pool in the watershed), N_{min} is the nitrogen mineralization of the fresh organic nitrogen pool (the amount of nitrogen moving from fresh organic, i.e., residue to the nitrate pool in the watershed), N_{atm} is the nitrate from atmospheric deposition, N_{denit} is the nitrate from denitrification, N_{leach} is nitrate percolation through the bottom layer of the soil profile in the watershed, N_{surf} is the nitrate loading to stream in the surface runoff in the watershed, and N_{latf} is the nitrate loading to stream in the lateral flow in the watershed. The unit of all the above terms is in kgN ha^{-1}.

Hass et al. [31] found that the nitrogen uptake distribution, or β_n, was strongly correlated with the nitrate concentration in crops. In the periods of increased nitrate uptake by plants in the root zone, the dominant phases of the nitrogen uptake distribution β_n were observed, which indicated that the crops consumed nitrate. SWAT calculates the nitrogen removed from the soil by plants by taking nitrogen from the nitrate pool [30]. If the nitrates in the upper layers of the soil was insufficient, the nitrates in the root zone were allowed to fully compensate for it. The actual amount of nitrogen removed from the soil, $N_{actualup,ly}$, is calculated using Equation (5):

$$N_{actualup,ly} = min \left[N_{up,ly} + N_{demand}, NO3_{ly} \right] \tag{5}$$

with:

$$N_{up,ly} = N_{up,zl} - N_{up,zu}$$

where $N_{up,ly}$ is the potential nitrogen uptake for layer *ly* (kgN ha^{-1}), N_{demand} is the nitrogen uptake demand not met by overlying soil layers (kgN ha^{-1}), $NO3_{ly}$ is the nitrate content of the soil layer *ly* (kgN ha^{-1}), $N_{up,zl}$ is the potential nitrogen uptake from the soil surface to the lower boundary of the soil layer (kgN ha^{-1}), and $N_{up,zu}$ is the potential nitrogen uptake from the soil surface to the upper boundary of the soil layer (kgN ha^{-1}).

To calculate the potential nitrate uptake $N_{up,z}$, from the soil surface to the depth z, Equation (6) is used:

$$N_{up,z} = \frac{N_{up}}{[1 - exp(-\beta_n)]}\left[1 - exp\left(-\beta_n \cdot \frac{z}{z_{root}}\right)\right] \tag{6}$$

with:

$$N_{up} = min\left[bio_{N,opt} - bio_N, \; 4fr_{N,3}\Delta bio\right]$$

and

$$bio_{N,opt} = fr_N bio$$

where N_{up} is the potential nitrogen uptake (kgN ha^{-1}), β_n is the nitrogen uptake distribution parameter, z is the depth from the soil surface (mm), z_{root} is the depth of root development into the soil (mm), $bio_{N,opt}$ is the optimal mass of nitrogen stored in plant material for the current growth stage (kgN ha^{-1}), bio_N is the actual mass of nitrogen stored in plant material (kgN ha^{-1}), fr_N is the normal fraction of nitrogen in the plant biomass, Δbio is the potential increase in total plant biomass on a given day (kg ha^{-1}), bio is the total plant biomass on a particular day (kg ha^{-1}), and subscript 3 indicates the maturity growth tage.

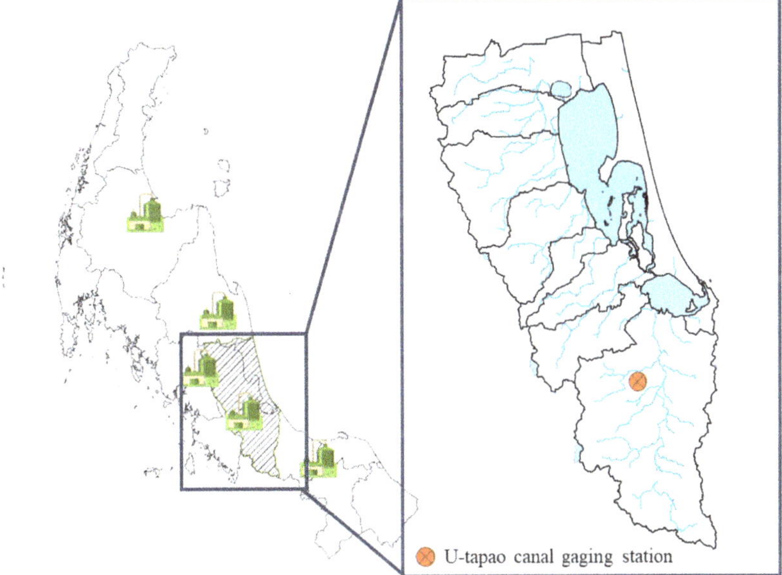

Figure 1. Map of the study area (Songkhla Lake Basin) in southern Thailand and the location of U-tapao canal gauging station.

2.3. Data Used

For ease of calculation, the watershed was divided into small units called HRUs. An HRU is the smallest spatial unit that consists of a unique load combination of land use, soil type, and slope. Figure 2 display the geographical data used in this study, including land use and elevation. To determine the slope, slope length, and stream network of each basin, digital elevation model (DEM) data with a resolution of 30 m was extracted from CGIAR CSI [32], as shown in Figure 2a. The topographical map thus obtained needed to be integrated with soil and land use maps to obtain HRUs in the area of interest. To integrate these data, all maps were converted to a raster dataset with a resolution of 50 m. The spatial distribution of soil types was provided by the FAO-UNESCO harmonized world soil database [33]. Figure 2b displays the land use map from the Land Development Department (LDD) of Thailand used in this study. To calculate plant growth and water and nitrate cycles occurring in the watershed, weather data from the National Centers for Environmental Prediction (NCEP) called Climate Forecast System Reanalysis (CFSR) [34] were used, which was recommended by SWAT developers and various studies utilized to predict crop production, streamflow, sediment yield, and nitrate load [35–37]. These data include the maximum and minimum temperatures, precipitation, solar radiation, and wind speed. Since the data from NCEP were available only from 1979 to 2014, a daily weather generator algorithm (dGEN) of the SWAT model was used to generate weather data during the rest of the calculation period. The CFSR data are provided on a Gaussian grid, defined by the NCEP, with a horizontal resolution of 38 km (0.3125°); the vertical resolution was not equally spaced. By combining the abovementioned data, we obtained ready-to-use data of the watershed of interest.

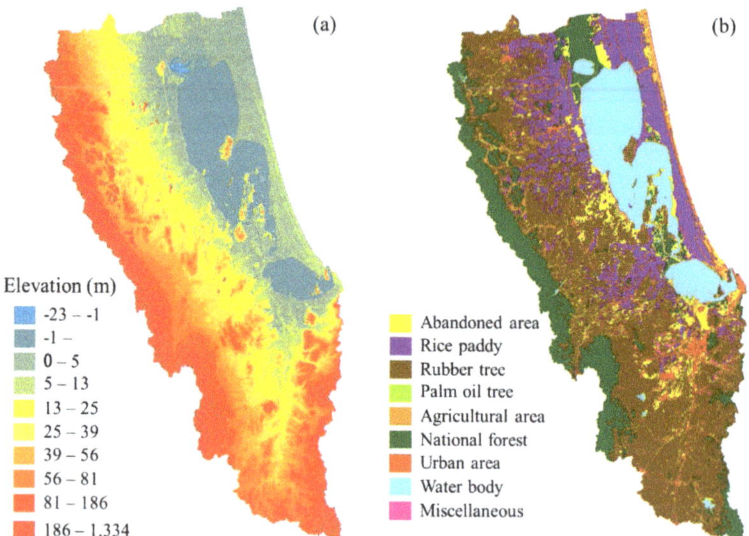

Figure 2. Maps of (**a**) elevation and (**b**) land use in Songkhla Lake Basin.

2.4. Model Calibration and Validation

To obtain a reliable prediction, the model must be carefully calibrated and validated. The main focus of this study was to investigate the effects of Napier grass plantations on hydrology and water quality; hence, streamflow and sediment yield observations were used for the calibration and validation. Since weather data greatly affected the simulation outputs, and since it is unclear if the weather data, including the maximum and minimum temperature, precipitation, solar radiation, and wind speed, generated by dGEN resembled the actual historical data, precipitation was also included in the calibration and validation processes. Only precipitation was selected for the process because it significantly affected the simulation outputs of the hydrological model, and because precipitation observation data were available. Although there are several gauging stations in the Songkhla basin, data on the hydrology, water quality, and precipitation of most stations are not publicly available. To the best of our knowledge, the only station that can be readily accessed for streamflow, sediment yield, and precipitation data is the U-tapao canal gaging station (6°55′52.32″ N, 100°26′24.72″ E, see Figure 1). Therefore, the monthly streamflow, sediment yield, and precipitation data from the station during 2009–2018 were used for calibration and validation.

For simulation using the SWAT model, there were several parameters affecting hydrology, water quality, and precipitation, and approximately 13 parameters exist that are related to the output of interest. With such a large number of parameters, it is difficult to perform manual calibration. Therefore, to perform the calibration, four steps were used to adjust the parameters. In the first step, previous studies [22,23,38,39] were reviewed to identify a range of parameters and sensitive parameters. Then, as a starting point, a simulation was performed using the default values suggested by the SWAT. Subsequently, the sensitive parameters were calibrated manually, similar to the manual calibration in Mengistu et al. [40] and Arnold et al. [41], except for the curve number that dos Santos et al. [22] suggested for multiplying the default numbers by 0.7. Finally, when needed, the input parameters were re-adjusted within reasonable parameter ranges obtained from the first step, and the process was repeated until satisfactory results were obtained. The coefficient of determination (R^2) of streamflow, sediment yield, and precipitation was used as an objective function, and the criterion for judging the quality of calibration was to identify the set of parameters that improved R^2 of all outputs to the desirable value of 0.70. The calibrated values are presented in Table 1. After a set of reliable parameters was obtained, the SWAT model was validated using three statistical parameters: the Nash–Sutcliffe model efficiency index (NSE), percent bias (PBIAS), and R^2.

Table 1. Summary of calibrated SWAT parameters.

	Parameter	File	Description (Unit)	Default Range	Previous Studies [33–36]	Default Value	Calibrated Value
Hydrology	CN2	.mgt	SCS runoff curve number	0–100	Default ×0.7	depends on soil and land use	55–69
	ESCO	.hru	Soil evaporation compensation factor	0–1	0.6–0.9	1	0.9
	CANMX	.hru	Maximum canopy storage (mm)	0–100	15–80	0	20
	ALPHA_BF	.gw	Baseflow recession constant	0.0071–0.0161	0.01–0.048	0.048	0.048
	GW_REVAP	.gw	Ground water revap coefficient	0–0.4	0.13–0.04	0.003	0.04
	GW_DELAY	.gw	Ground water delay (days)	0–500	14–500	31	14
	REVAPMN	.gw	Threshold depth of water in shallow aquifer for revap to occur	0–1000	250–500	750	500
	EVRCH	.bsn	Reach evaporation adjustment factor	0.5–1	0.5–0.9	1	0.9
	SURLAG	.bsn	Surface runoff lag coefficient	0.05–24	15	4	15
Sediment	SPCON	.bsn	Linear parameter for calculating the maximum amount of sediment that can be re-entrained during channel sediment routing	0.0001–0.01	0.001–0.008	0.0001	0.001
	LAT_SED	.hru	Sediment concentration in lateral flow and groundwater flow	0–5000	5.7–3000	0	3000
	CH_COV1	.rte	Channel erodibility factors	0–0.6	0.1–0.17	0	0.1
	CH_COV2	.rte	Channel cover factors	0–1	0.1–0.6	0	0.6
Precipitation	rexp	-	The exponent of the exponential distribution	1.0–2.0	-	-	1.3

2.5. Napier Grass Plantation Cases and Calculation Setting

The land use in the SLB at present (i.e., the baseline) and in case of Napier grass plantation are shown in Table 2. The non-hunting area must be preserved, and thus cannot be used for planting Napier grass. To avoid conflicts with existing industrial, urban, economical, and agricultural lands, only abandoned areas were considered for Napier grass plantations It should be noted that land used for agricultural purposes was found to decrease because those areas were considered abandoned agricultural lands by the LDD.

Table 2. The area of land use types in the baseline case, which is current agricultural land use, and the cases where the abandoned areas were utilized to plant Napier grass in Songkhla Lake Basin.

Land Use Type	Baseline Case		Napier Grass Case		%Change
	Area (ha)	%	Area (ha)	%	
Rice	118,769.9	14.6	113,769.9	13.9	−4.2
Rubber Trees	347,112.5	42.6	337,112.5	41.3	−2.9
Oil Palm	5592.5	0.7	5592.5	0.7	0.0
Agricultural Land	31,624.8	3.9	28,624.8	3.5	−9.5
Forest—Mixed	112,485.4	13.8	112,485.4	13.8	0.0
Residential—Med/Low Density	46,483.2	5.7	46,483.2	5.7	0.0
Water	110,469.8	13.5	110,469.8	13.5	0.0
Miscellaneous area	2394.9	0.3	2394.9	0.3	0.0
Abandoned area	40,800.5	5.0	0	0.0	−100.0
Napier grass	0.0	0.0	58,800.5	7.2	–
Total area	815,733.5	100.0	815,733.5	100.0	

Since the growth of Napier grass is highly dependent on management practices [42], particularly on the amount of nitrogen fertilizer applied, a total of four cases were formulated, in each of which the amount of the applied nitrogen fertilizer varied from 0 to 500 kgN ha^{-1}, at four levels. These values were obtained from studies by the Animal Nutrition Division, the Department of Livestock Development, and the Ministry of Agriculture and Cooperatives, and were published between 1993 and 2005 [43–51]. It was found that nitrogen fertilizer could increase Napier grass DMY by up to three times the DMY without the fertilizer [43–51]. However, it is unclear if such a large amount of fertilizer negatively impacts the environment in any way. The purpose of this variation was to investigate the impact of such an intensive fertilizer. These cases were applied to the ready-to-use data for the watershed of interest (as described above) to evaluate the impacts of different nitrogen fertilizer levels on Napier grass DMY, streamflow, sediment yield, and nitrate load.

In this study, the calibration period was from 2009 to 2013, while data from 2014 to 2018 were used for validation. Five warm-up years were used in the model initialization, as suggested by Tudose et al. [52], and the investigation was carried out over 10 years. Since the preset parameters for Napier grass plantation did not exist in the original SWAT model, a parameter set for predicting Napier grass crop yield must be developed. The simulation setup, model calibration, and validation were described in our previous study [26] and will not be repeated here. Since the models for other land uses were well established, the default setups for each land use suggested by SWAT were applied, except for the abandoned and miscellaneous areas that were not defined in SWAT. These areas were assumed to have a low agricultural area based on but the fact that little agricultural activity has been observed in the area. Changes between the parameter set of abandoned land and Napier grass cultivation are listed in Table 3.

Table 3. SWAT model parameters for abandoned land and Napier grass cultivation.

Category	Parameter	Definition	Abandoned Land	Napier Grass
Land cover/plant	IDC	Land cover/plant classification	6 (perennial)	6 (perennial)
	BIO_E	Radiation use efficiency	30	38
	CHTMX	Maximum canopy height (m)	0.9	2.5
	RDMX	Maximum root depth (m)	1.3	2.2
Runoff	CANMX	Maximum canopy storage (mm)	20 (calibrated)	20
	CN	Curve number	65 (calibrated)	55
Sediment	USLE_C	Minimum USLE crop factor	0.003	0.003
Fertilizer	FMINN	Fraction of mineral N (NO_3 and NH_4) in fertilizer (kg min-N/kg fertilizer)	0	0.46
	FORGN	Fraction of organic N in fertilizer (kg org-N/kg fertilizer)	0	0
	FNH3N	Fraction of mineral N in fertilizer applied as ammonia (kg NH_3-N/kg fertilizer)	0	0

2.6. Multidisciplinary Assessment Supporting Decision Making for Utilizing Napier Grass

Since solely considering the impact of Napier grass plantations on land is insufficient for decision making, this study considers impacts such as energy supply, carbon reduction, and benefits to farmers, in addition from hydrological impacts, in order to provide a better overview for decision making. From the simulation results obtained from the SWAT model, Napier grass DMY was utilized to evaluate energy supply, carbon reduction, and farmer benefits.

For energy supply, based on our previous study [27], approximately 11.1 kt-DMY was required for generating 1 MW of electricity. This is based on the assumption that a methane yield of 242 m^3 can be obtained from 1 ton of Napier grass DMY [9]. In addition, a methane energy density of 40 MJ m^{-3} [53] and an energy conversion efficiency of 30% was assumed; the potential power generation could be conveniently evaluated using a factor of 11.1 kt-DMY MW^{-1}. Beyond the benefit of obtaining electricity, Napier-grass-based power generation could serve as a substitute for conventional power generation derived from fossil fuels. In a previous study [54], by utilizing Napier-grass-derived natural gas for electric generation instead of fossil fuels, approximately 60% of CO_2 emissions could be reduced (i.e., from 1080 to 450 $kgCO_2$ MWh^{-1}). Therefore, the CO_2 reductions were estimated by multiplying the derived power generation by the reduced CO_2 emissions.

Using Napier grass as a biogas feedstock not only helps reduce CO_2 emissions but also provides benefits to farmers. Currently, the Napier grass purchase price in Thailand is approximately 300 Baht (t-fresh biomass) $^{-1}$ (equivalent to 1500 Baht t-DMY^{-1}). Furthermore, the cost of nitrogen fertilizers was only approximately 30 Baht kgN^{-1} at the time of this study. This is a relatively high purchase price with a relatively low additional cost and, thus, it was encouraging for farmers to aim to achieve more production per area. This could lead to environmental problems owing to the overutilization of nitrogen fertilizers. Therefore, the tradeoff between Napier grass production and nitrate loads should be carefully considered. The main purpose of this work is to investigate the impact of different applied fertilizer inputs on additional revenue. The money spent on fertilizers was deducted from the Napier grass selling price to evaluate the farmer's operating income under different management practices.

After all impacts, including surface runoff, sediment yield, nitrate load, energy supply, carbon reduction, and benefits to farmers, were determined in each case, they were scaled using max–min normalization to make it convenient for comparison. The max and min values were set to 1.0 and 0.5, respectively. The comparison was performed using a radar chart to enhance visibility.

3. Results and Discussion

3.1. Model Calibration and Validation

Figure 3 show the goodness-of-fit plots for the monthly streamflow, sediment yield, and precipitation, with the initial and final parameter sets in the calibration period from 2009 to 2013. Using the manual calibration process mentioned above, the R^2 of the streamflow increased from 0.476 to 0.714, as depicted in Figure 3a. For the sediment yield, the R^2 of 0.828 from the default parameter set was improved to 0.957 (see Figure 3b). The initial parameter set provided a reasonable prediction for precipitation, as an R^2 of 0.476 was initially achieved. However, the accuracy can be further improved after calibration, and an R^2 of 0.806 was obtained. It can be seen that the data on the top right in the F are quite far from most of the values presented in the figure. This is due to the fact that, in the wet season, the amount of precipitation is usually higher as compared to the rest of the year. Although they seem to be outliers in Figure 3, considering all the data used in the study (2005–2018, including the warm-up years), these events occurred once in a while and are normally found. It should be noted that some points are over/underpredicted and shifted as the model predicted that the event would occur one month before or after the actual event (see Figure 1). However, since the objective of this study is to estimate the long-term impacts of Napier grass plantation, the annual average of the results is sufficient for the estimation. Overall, it is clear that predictions can be satisfactorily improved by using the manual calibration process; Table 1 displays the final parameter set.

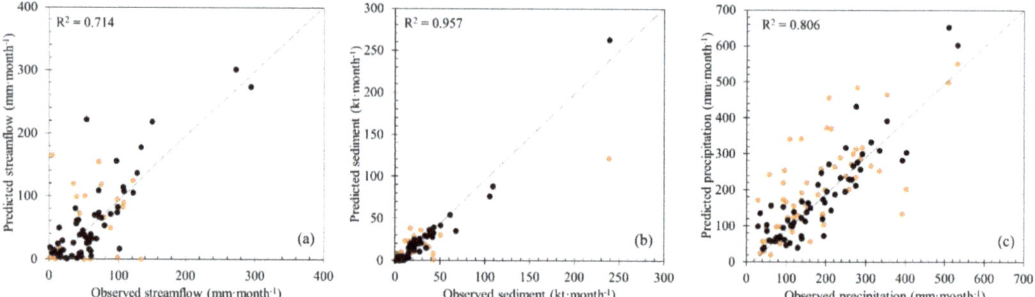

Figure 3. Goodness-of-fit plots with 1:1 line comparing the observed and simulated (**a**) streamflow, (**b**) sediment, and (**c**) precipitation with parameters before and after calibration during 2009–2013 at U-tapao canal gaging station. Presented R^2 is after calibration, where the orange (•) and black dots (•) indicate the predicted data before and after calibration, respectively.

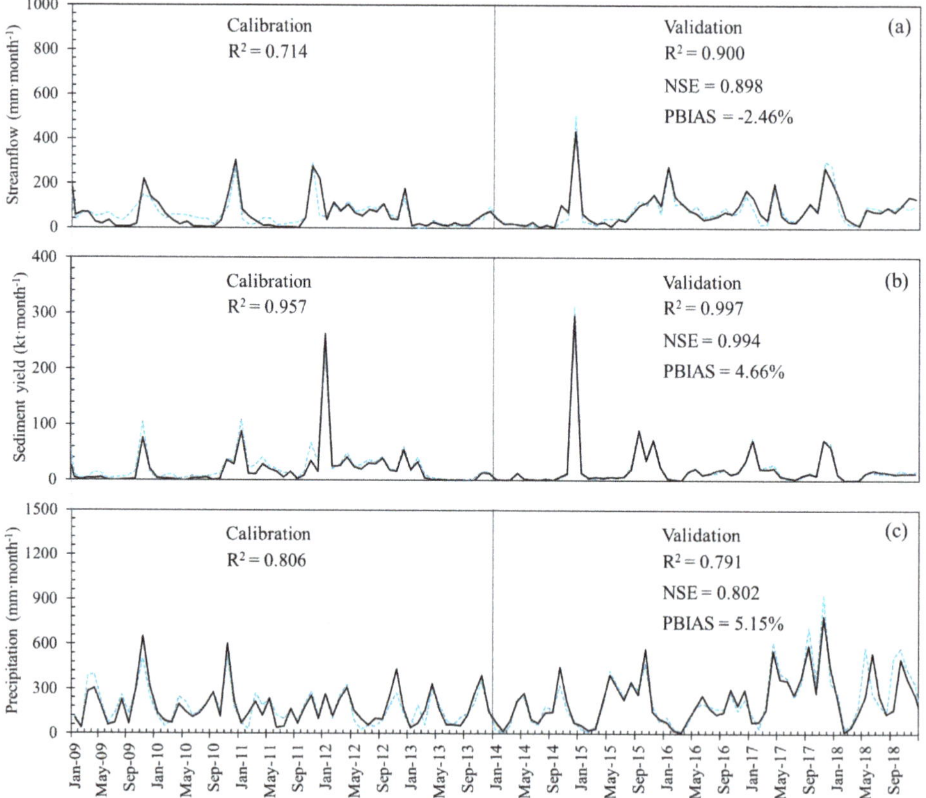

Figure 4. Model calibration and validation results for (**a**) streamflow, (**b**) sediment, and (**c**) precipitation at Songkhla Lake Basin at U-tapao canal gauging station, where the black solid line indicates the predicted data, and the blue dash line indicates the observed data.

To validate the generalization performance of the calibrated model, data from 2014 to 2018 were compared with the simulation data obtained from the parameter set obtained after the calibration. Figure 4 compares the temporal changes in the simulated and observed monthly streamflow, sediment yield, and precipitation obtained during calibration and validation. The results reveal that although some parts were over/underestimated, the model could reasonably predict the overall variation in streamflow, sediment yield, and precipitation. There was a slight concern regarding the accuracy of the weather data generated by the dGEN because precipitation data are generally recognized as the most important data for hydrological analysis. To ensure that the data were of adequate quality, careful validation was performed. It is clear that the precipitation data generated by dGEN resembled the actual historical data, as the dGEN could predict the precipitation data from 2013–2018, with an R^2 of 0.791, an NSE of 0.802, and a PBIAS of 5.15%.

With accurate weather data, it was found that the SWAT model can successfully and accurately estimate the streamflow during 2014–2018, with an R^2 of 0.900, an NSE of 0.898, and a PBIAS of −2.46%. The negative value of the PBAIS indicates that the model overestimated the streamflow by approximately 2.5% (on average). On the other hand, the sediment yield at the U-tapao canal gaging station during 2014–2018 can be estimated by the SWAT model with an R^2 of 0.997, an NSE of 0.994, and a PBIAS of 4.66%. The sediment yield was positive for the PBAIS, indicating that sediment yield was underestimated by approximately 4.7%. Considering all the statistical indicators, the model is deemed adequate for investigating the effects of Napier grass plantations on hydrology and water quality.

3.2. Impacts of Napier Grass Energy Plantation Cases

The impacts of different levels of applied nitrogen fertilizer on Napier grass production, streamflow, sediment yield, and nitrate load were investigated over a period of 10 years. Figure 5 depict the spatial distribution of average Napier grass DMY planted with different nitrogen fertilizer levels in the abandoned area in SLB. The results, as shown in Figure 6, revealed that without applying the nitrogen fertilizer, the average DMY in the basin was approximately 11.28 t-DMY ha^{-1} (i.e., 663 kt-DMY in total); however, as the amount of applied nitrogen fertilizer increased, the DMY increased. The Napier grass DMY can be increased to 18.19, 22.71, and 27.52 t-DMY ha^{-1} after the application of nitrogen fertilizers of 125, 250, and 500 kgN ha^{-1}, respectively. These results align with the hypothesis from Hazary et al. [42] that the fertilizer application rate positively affects the production of dedicated energy crops. The DMY increased by approximately 61% when the nitrogen fertilizer of 125 kgN ha^{-1} was applied; however, when the amount of fertilizer was doubled to 250 kgN ha^{-1}, only a 25% increase in DMY was observed. An increase in the DMY of only 22% was achieved when the nitrogen fertilizer level was further increased to 500 kgN ha^{-1}. Considering the diminishing return, it is unsurprising that a fertilizer level of 250 kgN ha^{-1} was recommended by the handbook from Nakhon Ratchasima Animal Nutrition Research and Development Center [55]. Although DMY was increased along with the amount of nitrogen fertilizer, it was not clear how it affected hydrology and water quality. Hence, it is important to investigate its effects on surface runoff, sediment yield, and nitrate load.

Figure 7 displays the simulated surface runoff, sediment load, and nitrate load obtained from the SWAT model. The surface runoff at the SLB for different cases was investigated, as shown in Figure 7a. It is clear that planting Napier grass in abandoned areas has a positive impact on surface runoff prevention. While reducing surface runoff may be beneficial for flood control, it can be considered detrimental for water resources and lake ecosystem health. This is due to the fact that the SLB is extremely prone to flooding and landslides. Thus, the decrease in surface runoff was considered to have a positive effect on the area. Surface runoff can be reduced by approximately 30% by Napier grass plantations. These results concur with the results of previous studies, that show that perennials can help reduce surface runoff [24,56]. This is due to the fact that most studies replaced row crops with perennials, and the perennials have better soil cover. In this study, abandoned areas were used in Napier grass plantations; based on Equations (1) and (2), the surface runoff Q_{surf} is a function of the curve number. The curve number directly reflects the characteristic land cover and hydrologic soil groups. When the abandoned areas were replaced by Napier grass, the curve number decreased from 65 to 55, resulting in better water retention. The lower curve number is likely due to the large transpiration rate of the Napier grass. While the surface runoff greatly decreased with the Napier grass plantation, no significant differences between the case of different applied nitrogen fertilizer levels were observed. This is because the increase in the vertical growth of Napier grass did not affect the lateral soil coverage, which is a key factor in reducing surface runoff [28].

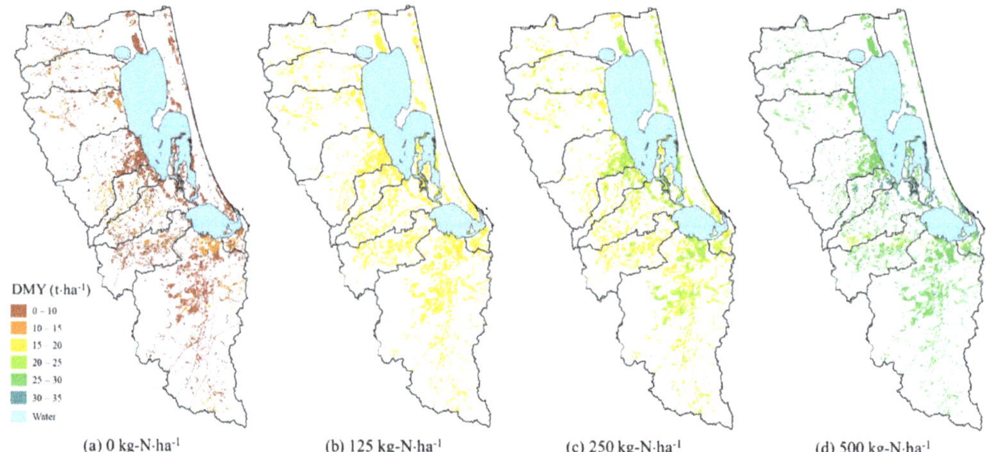

Figure 5. Spatial distribution of Napier grass DMY from abandoned areas in Songkhla Lake Basin under different nitrogen fertilizer application levels of (**a**) 0, (**b**) 125, (**c**) 250, and (**d**) 500 kgN ha^{-1}.

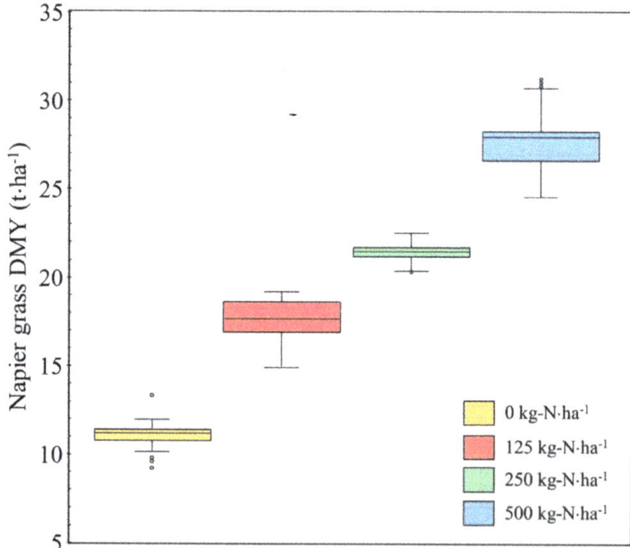

Figure 6. Box plots of Napier grass DMY from abandoned areas in Songkhla Lake Basin under different nitrogen fertilizers of 0, 125, 250, and 500 kgN ha^{-1}.

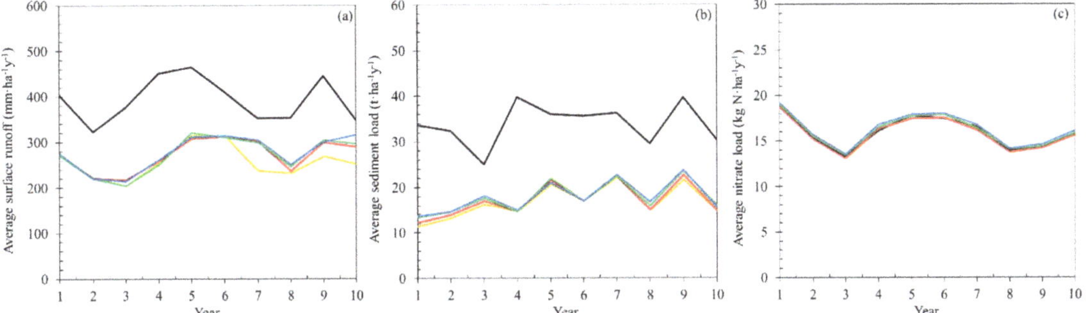

Figure 7. Average annual (**a**) surface runoff, (**b**) sediment load, and (**c**) nitrate load of Songkhla Lake Basin for 10 years timespan, where black, yellow, red, green, and blue solid lines indicate the baseline case and the Napier grass plantation cases with the nitrogen fertilizer applications of 0, 125, 250, and 500 kgN ha^{-1}, respectively.

Figure 7b displays the sediment yield for the different planting scenarios at the SLB. The sediment yield decreased when abandoned areas were used for Napier grass plantations. This has a positive impact, as the sediment yield is strongly related to soil erosion. The decrease in sediment yields implied that water bodies would be less polluted by soil erosion; sediment yield was significantly reduced, by approximately 50%. When different nitrogen fertilizer levels were applied, no significant differences were observed. These results were similar to those of the surface runoff presented above. It is unsurprising that the SWAT model utilizes the MUSLE equation (see Equation (3)), where the surface runoff volume and peak flow rate were used while calculating sediment yield. In addition to these two factors, the USLE cover and management factor C are the only parameters that change with land use, which involves only the impacts of crop type and tillage method. Since the applied fertilizer level did not affect the USLE cover and management factor C, it is unsurprising that the sediment yield was not affected by the different fertilizer levels. It is worth noting that the USLE_C, which is the minimum USLE crop factor, is the same for abandoned areas as well as Napier grass plantations, as shown in Table 3. This is because the crop types considered in the abandoned and Napier grass plantation areas were the same. A USLE_C of 0.003 was suggested as a default value for perennials; however, the USLE cover and management factor C could be different because it was calculated based on the USLE_C by considering the seasonal effects. Moreover, Singh et al. [57] suggested that the USLE cover and management factor C are the least influential parameters in sediment yield calculation.

For the nitrate loads calculated as the sum of leaching and loading to the water stream by surface runoff and lateral flow, the average nitrate loads over SLB with different planting cases are shown in Figure 7c. The results revealed that the nitrate loads can be reduced slightly when fertilizer rates of 0 and 125 kgN ha^{-1} were applied. The reduction in nitrate loads is in agreement with the results of previous studies, which indicated that dedicated energy crops consume much nitrogen for growth [42]. In addition, several studies have shown that perennials can help reduce nitrate loads [24,25,58,59]; however, the nitrate loads increased slightly when the amount of applied nitrogen fertilizer exceeded 250 kgN ha^{-1}. Although a large amount of nitrogen fertilizer was applied in the cultivation, the nitrate loads, as compared to the baseline (see above), increased by approximately 1.13% and 2.32% for the applied fertilizer rates of 250 kgN ha^{-1} and 500 kgN ha^{-1}, respectively. This can be explained by the total nitrate balance summarized in Table 4. It is clear that nitrogen uptake by plants was the most influential nitrogen removal process. Because of diminishing returns, Napier grass cannot consume all the applied fertilizer for the case of 250 and 500 kgN ha^{-1}, resulting in surplus nitrogen in the considered area.

Table 4. Soil system nitrate balance of the baseline and Napier grass plantation cases with four fertilizer application levels in the SLB. Values are expressed per hectare of the whole basin (kg-N ha^{-1} y^{-1}), including all land uses in the basin.

Item	Baseline	Napier Grass Plantation			
		0 kgN ha^{-1}	125 kgN ha^{-1}	250 kgN ha^{-1}	500 kgN ha^{-1}
Inputs					
Fertilizer application	39.88	39.88	42.90	47.15	51.88
Humus mineralization	9.36	9.06	9.31	10.15	10.31
Residue mineralization	6.78	6.37	6.76	8.12	8.94
Atmospheric deposition	0.26	0.26	0.26	0.26	0.26
$\sum$ Inputs	56.28	55.58	59.23	65.68	71.39
Outputs					
Denitrification	3.65	3.65	4.00	5.65	6.44
Nitrate uptake	37.64	37.84	40.37	43.64	47.51
Nitrate leached	13.86	13.34	13.86	14.53	14.58
Nitrate loading to stream in surface runoff	1.19	1.16	1.20	1.29	1.38
Nitrate loading to stream in lateral flow	0.04	0.04	0.04	0.04	0.04
$\sum$ Outputs	56.39	56.03	59.47	65.15	69.95
$\sum$ Inputs − $\sum$ Outputs	−0.11	−0.45	−0.24	0.53	1.44

To obtain a better basis for decision making, a multidisciplinary evaluation was carried out to compare the advantages and disadvantages of different planting cases. Figure 8 shows the radar chart of the evaluation indicators, including surface runoff, sediment yield, nitrate load, energy supply, farmer income, and CO_2 reduction for different planting cases. The results revealed that although applying nitrogen fertilizers of 500 kgN ha^{-1} provided the highest benefits in energy supply, farmer's incomes, and CO_2 reduction, it also performed the worst in hydrological indicators among the different planting cases considered in this study. Together with the case of 250 kgN ha^{-1} nitrogen fertilization, these were the only two cases that performed worse than the baseline upon increasing the amount of nitrate load into the system. On the other hand, without the applied fertilizer, benefits from Napier grass were in contrast with that of the case when nitrogen fertilization of 500 kgN ha^{-1} was applied. This suggests that there is a trade-off between hydrological indicators and other factors, including energy supply, farmer income, and CO_2 reduction. The case in which nitrogen fertilization of 125 kgN ha^{-1} was applied would be a better choice as it was more balanced in all indicators.

Overall, from the simulated results of this study, Napier grass plantation in the abandoned land in SLB resulted in a decrease in surface runoff and sediment yield, which is beneficial to the water cycle control in SLB since the SLB is prone to flooding and landslides. In addition, nitrate loads were shown to be reduced in the Napier grass plantation cases with modest fertilizer applications. The socio-economic indicators supported utilizing abandoned areas in southern Thailand to plant Napier grass for biogas-based power generation, which can help reduce the dependency on imported electricity and provide additional income and/or job opportunities for local people. However, it should be noted that the decrease in surface runoff, sediment yield, and nitrate load does not always have a positive impact on ecosystem health in areas that are not susceptible to flooding and landslides. Therefore, prior to the introduction of new dedicated energy crops, it is important to assess the impacts on land, ecosystems, and other criteria unique to the area of interest. Although there are several potential benefits to be obtained from Napier grass plantations, it is unclear if the Napier grass-related businesses will be economically sustainable. In this study, the analysis was not quantitative because the importance of all evaluation indicators could not be adequately compared. Therefore, a further study on the economic perspective of introducing Napier grass as a biogas feedstock for power generation should be carried out; such a study is already ongoing within our research group.

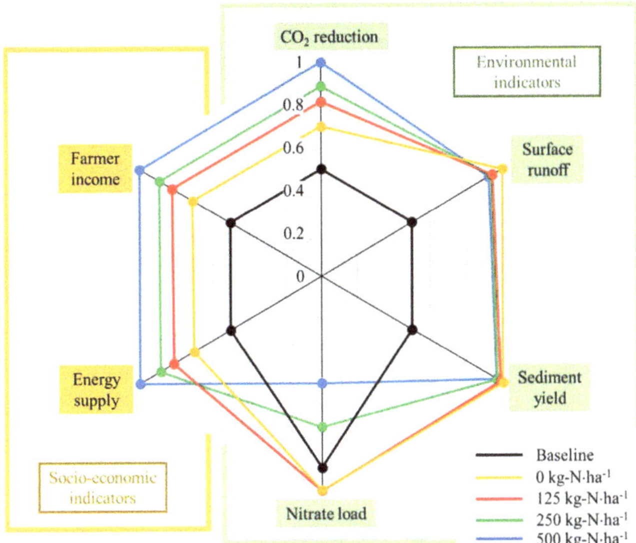

Figure 8. Comparison of the evaluation indicators of baseline and Napier grass plantation cases on surface runoff, sediment yield, nitrate load, energy supply, farmer income, and CO_2 reduction. The indicators were scaled by the max–min normalization of the values, where max and min values were set to be 1.0 and 0.5, respectively.

4. Conclusions

To introduce new crops for specific purposes, such as bio-energy resources, it is important to consider their impacts on environmental and socioeconomic benefits. In this study, a methodological framework for investigating the impacts of Napier grass plantations and a multidisciplinary assessment was successfully developed based on the SWAT watershed model. To obtain a reliable parameter set for the simulation, this model was carefully calibrated and validated. Utilizing manual calibration, a set of parameters used to predict streamflow, sediment yield, and nitrate load were obtained by considering the R^2. The results showed that by planting Napier grass, surface runoff, sediment yield, and nitrate load can all be greatly reduced. This is because of the increase in land cover and the nature of Napier grass, which consumes a large amount of nitrogen. The increase in nitrogen fertilizer was found to be relatively insignificant to overall surface runoff and sediment yield; however, the amount of N fertilizer significantly affected the nitrate load—as the nitrogen fertilizer level increased, the nitrate load increased. To have a clearer idea of how different cases impacted other perspectives, energy supply, farmer's incomes, and CO_2 reduction were included as further considerations. The results of this consideration revealed that when no fertilizers were applied, the management practice performed best in reducing the negative impacts on hydrology and water quality. However, applying fertilizer as high as 500 kgN ha^{-1} provided the highest energy supply, income to farmers, and CO_2 reduction. The results of this study provide information about the environmental impacts as well as crop production. This is supportive for both energy-related policymakers and farmers, since policymakers can utilize this information to consider a tradeoff between environmental impacts and crop production, and the farmers can learn how to achieve high comprehensive benefits from their crops.

Author Contributions: Conceptualization, K.N. and T.M.; methodology, K.N.; software, K.N.; validation, K.N.; formal analysis, K.N. and T.M.; investigation, K.N. and P.C.-a.; resources, K.N.; data curation, K.N.; writing—original draft preparation, K.N.; writing—review and editing, P.C.-a. and T.M.; visualization, P.C.-a.; supervision, T.M. and K.H., funding acquisition, K.H. All authors have read and agreed to the published version of the manuscript.

Funding: This work was carried out by the joint research program of the Institute of Materials and Systems for Sustainability, Nagoya University.

Acknowledgments: Gives thanks to the fund.

Conflicts of Interest: The authors declare no conflict of interest.

References

1. Höhn, J.; Lehtonen, E.; Rasi, S.; Rintala, J. A Geographical Information System (GIS) based methodology for determination of potential biomasses and sites for biogas plants in southern Finland. *Appl. Energy.* **2014**, *113*, 1–10. [CrossRef]
2. Chukwuma, E.C.; Okey-Onyesolu, F.C.; Ani, K.A.; Nwanna, E.C. Gis bio-waste assessment and suitability analysis for biogas power plant: A case study of Anambra state of Nigeria. *Renew. Energy* **2020**, *163*, 1182–1194. [CrossRef]
3. Piekutin, J.; Puchlik, M.; Haczykowski, M.; Dyczewska, K. The Efficiency of the Biogas Plant Operation Depending on the Substrate Used. *Energies* **2021**, *14*, 3157. [CrossRef]
4. Paine, L.K.; Peterson, T.L.; Undersander, D.J.; Rineer, K.C.; Bartelt, G.A.; Temple, S.A.; Sample, D.W.; Klemme, R.M. Some ecological and socio-economic considerations for biomass energy crop production. *Biomass Bioenergy* **1996**, *10*, 231–242. [CrossRef]
5. Shahpari, S.; Allison, J.; Harrison, M.T.; Stanley, R. An Integrated Economic, Environmental and Social Approach to Agricultural Land-Use Planning. *Land* **2021**, *10*, 364. [CrossRef]
6. Cervelli, E.; Scotto di Perta, E.; Pindozzi, S. Energy crops in marginal areas: Scenario-based assessment through ecosystem services, as support to sustainable development. *Ecol. Indic.* **2020**, *113*, 106180. [CrossRef]
7. Li, J.; Zhang, Z.; Jin, X.; Chen, J.; Zhang, S.; He, Z.; Li, S.; He, Z.; Zhang, H.; Xiao, H. Exploring the socioeconomic and ecological consequences of cash crop cultivation for policy implications. *Land Use Policy* **2018**, *76*, 46–57. [CrossRef]
8. Sinbuathong, N.; Sangsil, Y.; Sawanon, S. *Energy, Transportation and Global Warming*; Springer: Berlin/Heidelberg, Germany, 2016.
9. Sawasdee, V.; Pisutpaisal, N. Feasibility of Biogas Production from Napier Grass. *Energy Procedia* **2014**, *61*, 1229–1233. [CrossRef]
10. Dussadee, N.; Ramaraj, R.; Cheunbarn, T. Biotechnological application of sustainable biogas production through dry anaerobic digestion of Napier grass. *3 Biotech* **2017**, *7*, 47. [CrossRef]
11. Prapinagsorn, W.; Sittijunda, S.; Reungsang, A. Co-Digestion of Napier Grass and Its Silage with Cow Dung for Bio-Hydrogen and Methane Production by Two-Stage Anaerobic Digestion Process. *Energies* **2017**, *11*, 47. [CrossRef]
12. Bach, E.E.; Alcântara, V.B.G.; Alcântara, P.B.; Veasey, E.A. Biochemical and isoenzyme analyses of elephant grass, Pennisetum purpureum (Schum) varieties. *Sci. Agric.* **1995**, *52*, 528. [CrossRef]
13. Janejadkarn, A.; Chavalparit, O. Biogas Production from Napier Grass (Pak Chong 1) (Pennisetum purpureum × Pennisetum americanum). *Adv. Mat. Res.* **2013**, *856*, 327–332. [CrossRef]
14. Napier Grass. Available online: http://weben.dede.go.th/webmax/content/napier-grass (accessed on 5 September 2021).
15. Negawo, A.T.; Teshome, A.; Kumar, A.; Hanson, J.; Jones, C.S. Opportunities for Napier Grass (Pennisetum purpureum) Improvement Using Molecular Genetics. *Agronomy* **2017**, *7*, 28. [CrossRef]
16. Anderson, W.F.; Sarath, G.; Edme, S.; Casler, M.D.; Mitchell, R.B.; Tobias, C.M.; Hale, A.L.; Sattler, S.E.; Knoll, J.E. Dedicated Herbaceous Biomass Feedstock Genetics and Development. *BioEnergy Res.* **2006**, *9*, 399–411. [CrossRef]
17. Nelson, R.G.; Ascough, J.C.; Langemeier, M.R. Environmental and economic analysis of switchgrass production for water quality improvement in northeast Kansas. *J. Environ. Manag.* **2006**, *79*, 336–347. [CrossRef] [PubMed]
18. Pincam, T.; Brix, H.; Eller, F.; Jampeetong, A. Hybrid Napier grass as a candidate species for bio-energy in plant-based water treatment systems: Interactive effects of nitrogen and water depth. *Aquat. Bot.* **2017**, *138*, 82–91. [CrossRef]
19. Börjesson, P. Environmental effects of energy crop cultivation in Sweden—I: Identification and quantification. *Biomass Bioenergy* **1999**, *16*, 137–154. [CrossRef]
20. Njakou Djomo, S.; Ac, A.; Zenone, T.; De Groote, T.; Bergante, S.; Facciotto, G.; Sixto, H.; Ciria Ciria, P.; Weger, J.; Ceulemans, R. Energy performances of intensive and extensive short rotation cropping systems for woody biomass production in the EU. *Renew. Sust. Energy Rev.* **2015**, *41*, 845–854. [CrossRef]
21. Vatsanidou, A.; Kavalaris, C.; Fountas, S.; Katsoulas, N.; Gemtos, T. A Life Cycle Assessment of Biomass Production from Energy Crops in Crop Rotation Using Different Tillage System. *Sustainability* **2020**, *12*, 6978. [CrossRef]
22. Dos Santos, F.M.; Proença de Oliveira, R.; Augusto Di Lollo, J. Effects of Land Use Changes on Streamflow and Sediment Yield in Atibaia River Basin—SP, Brazil. *Water* **2020**, *12*, 1711. [CrossRef]
23. Epelde, A.M.; Cerro, I.; Sánchez-Pérez, J.M.; Sauvage, S.; Srinivasan, R.; Antigüedad, I. Application of the SWAT model to assess the impact of changes in agricultural management practices on water quality. *Hydrol. Sci. J.* **2015**, *60*, 825–843. [CrossRef]
24. Ng, T.L.; Eheart, J.W.; Cai, X.; Miguez, F. Modeling Miscanthus in the Soil and Water Assessment Tool (SWAT) to Simulate Its Water Quality Effects as a Bioenergy Crop. *Environ. Sci. Technol.* **2010**, *44*, 7138–7144. [CrossRef]
25. Cibin, R.; Trybula, E.; Chaubey, I.; Brouder, S.M.; Volenec, J.J. Watershed-scale impacts of bioenergy crops on hydrology and water quality using improved SWAT model. *Glob. Change Biol.* **2016**, *8*, 837–848. [CrossRef]
26. Nantasaksiri, K.; Charoen-Amornkitt, P.; Machimura, T. Land Potential Assessment of Napier Grass Plantation for Power Generation in Thailand Using SWAT Model. Model Validation and Parameter Calibration. *Energies* **2021**, *14*, 1326. [CrossRef]

27. Nantasaksiri, K.; Charoen-Amornkitt, P.; Machimura, T. Integration of multicriteria decision analysis and geographic information system for site suitability assessment of Napier grass-based biogas power plant in southern Thailand. *Renew. Sust. Energ.* **2021**, *1*, 100011. [CrossRef]

28. Somboonsuke, B.; Phitthayaphinant, P.; Sdoodee, S.; Kongmanee, C. Farmers' perceptions of impacts of climate variability on agriculture and adaptation strategies in Songkhla Lake basin. *Kasetsart J. Soc. Sci.* **2018**, *39*, 277–283. [CrossRef]

29. Cookey, P.E.; Darnswasdi, R.; Ratanachai, C. Local people's perceptions of Lake Basin water governance performance in Thailand. *Ocean. Coast. Manag.* **2016**, *120*, 11–28. [CrossRef]

30. Soil and Water Assessment Tool SWAT, Theoretical Documentation. Available online: https://swat.tamu.edu/media/1292/SWAT2005theory.pdf (accessed on 10 August 2021).

31. Hass, M.B.; Guse, B.; Pfannerstill, M.; Fohrer, N. Detection of dominant nitrate processes in ecohydrological modeling with temporal parameter sensitivity analysis. *Ecol. Model.* **2015**, *314*, 62–72. [CrossRef]

32. Farr, T.G.; Rosen, P.A.; Caro, E.; Crippen, R.; Duren, R.; Hensley, S.; Michael, K.; Mimi, P.; Ernesto, R.; Ladislav, R.; et al. The Shuttle Radar Topography Mission. *Rev. Geophys.* **2007**, *45*, RG2004. [CrossRef]

33. Fischer, G.; Nachtergaele, F.O.; Prieler, S.; Teixeira, E.; Tóth, G.; van Velthuizen, H.; Verelst, L.; Wiberg, D. *Global Agro-Ecological Zones (GAEZ v3.0): Model Documentation*; IIASA: Laxenburg, Austria; FAO: Rome, Italy, 2012; p. 179.

34. NCEP Climate Forecast System Reanalysis (CFRS). Available online: http://globalweather.tamu.edu/ (accessed on 23 August 2021).

35. Senent-Aparicio, K.; Jimeno-Sáez, P.; López-Ballesteros, A.; Ginés Giménez, J.; Pérez-Sánchez, J.; Cecilia, J.M.; Srinivasan, R. Impacts of swat weather generator statistics from high-resolution datasets on monthly streamflow simulation over Peninsular Spain. *J. Hydrol. Reg. Stud.* **2021**, *35*, 100826. [CrossRef]

36. Uniyal, B.; Dietrich, J.; Vu, N.Q.; Jha, M.K.; Luis Arumí, J. Simulation of regional irrigation requirement with SWAT in different agro-climatic zones driven by observed climate and two re-analysis datasets. *Sci. Total Environ.* **2019**, *649*, 846–865. [CrossRef]

37. Himanshu, S.K.; Pandey, A.; Yadav, B.; Gupta, A. Evaluation of best management practices for sediment and nutrient loss control using SWAT model. *Soil Tillage Res.* **2019**, *192*, 42–58. [CrossRef]

38. Ruan, H.; Zou, S.; Yang, D.; Wang, Y.; Yin, Z.; Lu, Z.; Li, F.; Xu, B. Runoff Simulation by SWAT Model Using High-Resolution Gridded Precipitation in the Upper Heihe River Basin, Northeastern Tibetan Plateau. *Water* **2017**, *9*, 866. [CrossRef]

39. Abbaspour, K.C.; Yang, J.; Maximov, I.; Siber, R.; Bogner, K.; Mieleitner, J.; Zobrist, J.; Srinivasan, R. Modelling hydrology and water quality in the pre-alpine/alpine Thur watershed using SWAT. *J. Hydrol.* **2007**, *333*, 413–430. [CrossRef]

40. Mengistu, A.G.; van Rensburg, L.D.; Woyessa, Y.E. Techniques for calibration and validation of SWAT model in data scarce arid and semi-arid catchments in South Africa. *J. Hydrol. Reg. Stud.* **2019**, *25*, 100621. [CrossRef]

41. Arnold, J.G.; Moriasi, D.N.; Gassman, P.W.; Abbaspour, K.C.; White, M.J.; Srinivasan, R.; Santhi, C.; Harmel, R.D.; van Griensven, A.; Van Liew, M.W.; et al. SWAT: Model use, calibration, and validation. *Trans. ASABE* **2012**, *55*, 1491–1508. [CrossRef]

42. Hazary, M.E.H.; Bilkis, T.; Khandaker, Z.H.; Akbar, M.A.; Khaleduzzaman, A.B.M. Effect of nitrogen and phosphorus fertilizer on yield and nutritional quality of Jumbo Grass (Sorghum Grass × Sudan Grass). *Adv. Anim. Vet. Sci.* **2015**, *3*, 444–450. [CrossRef]

43. Phaikeaw, C.; Phunphiphat, W.; Phunpiphat, R.; Kulna, S. Effect of Rate and Application time of Nitrogen Fertilizer on Forage Yield and Chemical Composition of Dwarf Napier Grass in Sukhothai Province. In *Annual Report of Bereau Animal Nutrition Development 2004*; Thai National AGRIS Centre: Bangkok, Thailand, 2004; pp. 45–54.

44. Vuthiprachumpai, L.; Nakamanee, G.; Punpipat, W.; Monchaikul, S. Effect of Plant Spacings on Yield and Chemical Com-position of Napier Grass (Pennisetum purpureum), Dwarf Elephant Grass (P. purpureum cv. Mott) and King Grass (P. pur-pureum x P. americanum) at Chainat Province. In *Annual Report of Bureau Animal Nutrition Development 1998*; Thai National AGRIS Centre: Bangkok, Thailand, 1998; pp. 194–228.

45. Suksaran, W.; Nuntachomchoun, P.; Vongpipatana, C. Yield and Cheimcal Compositions of Napier Grass in Various locations II Effect of Cutting Interval on Yield and Chemical Compositions of Three Varieties of Napier Grass (2.4) in Petchabun Province. In *Annual Report of Bureau Animal Nutrition Development 1991*; Thai National AGRIS Centre: Bangkok, Thailand, 1991; pp. 41–53.

46. Klum-Em, K.; Pojun, S.; Thammasal, P. Effect of Rate and Application Period of Nitrogen Fertilizer on Yield and Chemical Composition of Dwarf Napier Grass in Sa Kaeo Province. In *Annual Report of Bureau Animal Nutrition Development 2002*; Thai National AGRIS Centre: Bangkok, Thailand, 2002; pp. 159–174.

47. Waipanya, S.; Kulna, S.; Suriyachaiwatana, I.; Srichoo, C. Effect of Plant Spacing on Yield and Chemical Composition of 3 Varieties of Napier Grass in Nakhonpanom Province. In *Annual Report of Bureau Animal Nutrition Development 2003*; Thai National AGRIS Centre: Bangkok, Thailand, 2003; pp. 32–42.

48. Chainosaeng, W.; Nuntachomchoun, P.; Suksaran, W. Yield and Chemical Compositions of Napier Grass in Various Locations Effect of Plant Spacing on Yield and Chemical Compositions of Three Varieties of Napier Grass (1.4) in Petchabun Province. In *Annual Report of Bureau of Animal Nutrition Development 2004*; Thai National AGRIS Centre: Bangkok, Thailand, 2004; pp. 26–40.

49. Chubisaeng, W.; Bhokasawat, K.; Intarit, S.; Wongpipat, C. Effect of Rate and Application Time of Nitrogen Fertilizaer on Yield and Chemical Composition of Dwarf Napier Grass in Renu Soil Series. In *Annual Report of Bureau Animal Nutrition Development 2004*; Thai National AGRIS Centre: Bangkok, Thailand, 2004; pp. 55–78.

50. Yuthavoravit, C.; Suksaran, W.; Paotong, S. Effect of Rate and Application Time of Nitrogen Fertilizer on Yield and Chemical Composition of Dwarf Napier Grass in Hub-Kapong Soil Series. In *Annual Report of Bureau Animal Nutrition Development 2004*; Thai National AGRIS Centre: Bangkok, Thailand, 2004; pp. 79–102.

51. Phunphiphat, W.; Phunphiphat, R.; Kulna, S. Study on cost of production and forage yield of Dwarf Napeir Under Intensive Management. In *Annual Report of Bureau Animal Nutrition Development 2005*; Thai National AGRIS Centre: Bangkok, Thailand, 2005; pp. 87–99.

52. Tudose, N.C.; Marin, M.; Cheval, S.; Ungurean, C.; Davidescu, S.O.; Tudose, O.N.; Mihalache, A.L.; Davidescu, A.A. SWAT Model Adaptability to a Small Mountainous Forested Watershed in Central Romania. *Forests* **2021**, *12*, 860. [CrossRef]

53. Bossel, U.; Eliasson, B.; Taylor, G. The Future of the Hydrogen Economy: Bright or Bleak? *Distrib. Gener. Altern. Energy J.* **2003**, *18*, 29–70. [CrossRef]

54. Krittayakasem, P.; Patumsawad, S.; Garivait, S. Emission Inventory of Electricity Generation in Thailand. *JSEE* **2011**, *2*, 65–69.

55. Kiyothong, K. *Handbook of Napier Grass CV. Pakchong 1 Plantation*; Mittrapap Printing Ltd.: Nakhon Ratchasima, Thailand, 2011.

56. Dalzell, B.J.; Mulla, D.J. Perennial vegetation impacts on stream discharge and channel sources of sediment in the Minnesota River Basin. *J. Soil Water Conserv.* **2018**, *73*, 120–132. [CrossRef]

57. Singh, A.; Imtiyaz, M.; Isaac, R.K.; Denis, D.M. Assessing the performance and uncertainty analysis of the SWAT and RBNN models for simulation of sediment yield in the Nagwa watershed, India. *Hydrol. Sci. J.* **2014**, *59*, 351–364. [CrossRef]

58. Christianson, L.E.; Frankenberger, J.; Hay, C.; Helmers, M.J.; Sands, G. *Ten Ways to Reduce Nitrogen Loads from Drained Cropland in the Midwest*; University of Illinois Extension: Urbana, IL, USA, 2016.

59. Zhou, X.; Helmers, M.J.; Asbjornsen, H.; Kolka, R.; Tomer, M.D. Perennial Filter Strips Reduce Nitrate Levels in Soil and Shallow Groundwater after Grassland-to-Cropland Conversion. *J. Environ. Qual.* **2010**, *39*, 2006–2015. [CrossRef] [PubMed]

sustainability

Article

Development of Novel QAPEX Analysis System Using Open-Source GIS

Jayoung Koo [1], Jonggun Kim [2], Jicheol Ryu [3], Dong-Suk Shin [1], Seoro Lee [2], Min-Kyeong Kim [4], Jaehak Jeong [5] and Kyoung-Jae Lim [2,*]

[1] Watershed and Total Load Management Research Division, National Institute of Environmental Research, Incheon 22689, Korea; koo0105@naver.com (J.K.); sds8488@korea.kr (D.-S.S.)
[2] Department of Regional Infrastructure Engineering, Kangwon National University, Chuncheon 24341, Korea; kimjg23@gmail.com (J.K.); seorolee91@gmail.com (S.L.)
[3] Research Strategy and Planning Division, National Institute of Environmental Research, Incheon 22689, Korea; ryu0402@korea.kr
[4] Department of Agricultural Environment, National Institute of Agricultural Sciences, Wanju 55365, Korea; mkkim@rda.go.kr
[5] Blackland Extension and Research Center, Texas A&M AgriLife, College Station, TX 77843, USA; jjeong@brc.tamus.edu
* Correspondence: kyoungjaelim@gmail.com; Tel.: +82-033-250-6468

Abstract: The Agricultural Policy/Environmental eXtender (APEX) model has been used for farm/small watershed management, and the ArcAPEX interface was developed using the ArcGIS extension. However, the interface requires a paid license and limits dynamic applications that reflect various agricultural farming practices. In this study, a novel APEX model interface using Quantum GIS, the QAPEX analysis system, was developed by incorporating open-source-based GIS software for the simulation of water quality impacts of various best management practices reflecting local farming activities. The watershed delineation process running on the QAPEX interface is more flexible than that on the ArcAPEX interface, which renders simulations on hydrology and water quality with considerable precision. The newly developed system can be used to visually interpret simulation results (e.g., flow and load duration curve functions). Therefore, the open-source-based model can be used to derive data for sustainable agricultural policies, with a focus on the field-level application of management practices.

Keywords: APEX; open-source software; QGIS; QAPEX; best management practice; LDC

Citation: Koo, J.; Kim, J.; Ryu, J.; Shin, D.-S.; Lee, S.; Kim, M.-K.; Jeong, J.; Lim, K.-J. Development of Novel QAPEX Analysis System Using Open-Source GIS. *Sustainability* **2022**, *14*, 8199. https://doi.org/10.3390/su14138199

Academic Editor: Christian N. Madu

Received: 31 March 2022
Accepted: 30 May 2022
Published: 5 July 2022

Publisher's Note: MDPI stays neutral with regard to jurisdictional claims in published maps and institutional affiliations.

1. Introduction

An increase in air temperature, torrential heavy rainfall, and changes in rainfall patterns resulting from climate change affect the overall farming environment, including the rearing period and characteristics of crops [1]. In particular, concentrated heavy rainfall can increase the effect of water pollution from nonpoint pollution sources (NPSs) in agricultural regions, which necessitates the management of NPSs [2]. The Ministry of Environment (MOE) of South Korea has designated control areas of NPSs to reduce pollution from such sources, and is implementing numerous control measures and projects, such as establishing turbidity reduction measures and expanding NPS treatment facilities [3–5]. Furthermore, to establish control methods for NPSs, the MOE of South Korea has categorized 17 representative land covers through the Environmental Fundamental Data Examination project starting in 2008. Based on the findings from seven years of monitoring, the Ministry has determined the event mean concentration and NPS basic unit value by land cover, which provides basic data necessary for watershed management and watershed model operation [6,7].

The Soil and Water Assessment Tool (SWAT) [8] and Hydrological Simulation Program-FORTRAN [9] are used in watershed models in Korea to manage NPSs. In the Conservation Effects Assessment Project, a national project run by the Agricultural Research Service of the United States Department of Agriculture, Natural Resources Conservation Service, and Texas A&M AgriLife Research, the Agricultural Policy/Environmental eXtender (APEX) model and SWAT model were used to assess the effect of agricultural NPSs, and the Best Management Practices (BMPs) were applied to analyze conservation effects. Consequently, the findings have been used as national policy data for the United States [10].

The APEX model was developed with support from the USA Environmental Protection Agency (USEPA) to investigate the management effects of farming areas in small-scale watersheds. The model applies and evaluates BMPs because it can reflect various farming activities, including erosion from wind and water, economic feasibility, drainage for irrigation, intertilling, buffer strips, fertilizer and compost usage, crop rotation, pasturing, pesticide application, and plowing [11,12]. Furthermore, the Rural Development Administration (RDA) and Texas A&M AgriLife Research, the developer of the Environmental Policy Integrated Climate (EPIC)/APEX model, collaborated on the development of the APEX-Paddy model, which can be used to assess and control NPSs while considering the growing conditions of rice paddies, such as ponding water and transplanting [11,13]. The APEX model was improved into various models, such as WinAPEX [14], which is a basic APEX model with a simple and user-friendly interface; ArcAPEX [15], in which the ArcGIS software and APEX model are combined; and a web-based APEX model [16].

However, the WinAPEX system exhibits numerous shortcomings, such as the manual input of watershed characteristics, including soil properties, land use, slope, and waterway length, and searching through a long list of variables to modify them. In particular, the visual verification of the watershed information and simulation results of WinAPEX is required. Accordingly, the ArcAPEX interface improved on such shortcomings, enabled the automatic input and visualization of watershed information, and offered various geographic information system (GIS) functions. Nonetheless, the interface does not provide a visualization function for interpreting simulation results, and users are required to manually search and modify the necessary variables when applying BMPs in watersheds. Moreover, because the ArcGIS software, which is the base of the ArcAPEX interface, is a commercial program, users must purchase an expensive license. Because of the high cost of acquiring a license, the ArcGIS software is not widely used, and institutions such as the National Disaster Management Institute (NDMI) have substituted it with SuperMap (www.supermap.com, accessed on 1 February 2018) and Quantum GIS (QGIS) (www.qgis.org, accessed on 1 February 2018) [17].

Chen et al. [18] analyzed the functions of 31 open-source GIS software and revealed that, as an open-source GIS software, QGIS superior and the most appropriate option for use in the water resource field. In addition to data visualization, editing, and analysis functions, QGIS is capable of running most tasks available on commercial software. QGIS can be combined with other open-source GIS packages, such as PostGIS, GRASS, and MapServer, boosting its utility in various areas [18,19]. Furthermore, QGIS can be run on a number of operating systems, including Mac OS X, Linux, and Microsoft Windows.

This study linked the APEX model with open-source-based GIS software to develop the QAPEX analysis system, which includes the functions of the automatic input of watershed information, BMP-applicable interface, and visualization of simulation results.

The novelty of this study is the development of a user-friendly QAPEX analysis system for agricultural nonpoint pollution prediction and BMP evaluation by linking the APEX model and open-source-based GIS software for the first time. This QAPEX system provides multiple land use and soil combinations, which was not possible with previous WinAPEX and ArcAPEX interfaces, to provide an accurate representation of various land uses within the subbasin. Furthermore, the QAPEX system provides Flow Duration Curve (FDC)/Load–Duration Curve (LDC) analysis functions to be directly used for watershed management by considering flow regimes and water quality/quantity. The QAPEX analysis system developed in this study can quantify nonpoint pollution loads on agricultural lands based on a field scale and evaluate the effects of agricultural nonpoint pollution reduction for various structural and nonstructural BMPs. Furthermore, from the perspective of sustainable agricultural hydrological environment management, QAPEX could be used as an environmental tool to support policy decision-making necessary for various environmental impact assessments and strategic management plans.

2. Materials and Methods

2.1. Description of APEX Model and BMPs

Factors for operating lands in the APEX model originate from the EPIC model [20], which was developed in the early 1980s to evaluate the impact of erosion on productivity. The drainage area considered in the EPIC model is generally set at a maximum of 100 ha under the condition that the weather, soil, and management system remain constant [12]. The main components of the EPIC model are weather simulation, floodgate, erosion–sedimentation, nutrient cycling, decomposition of pesticides, crop growth, soil temperature, cultivation, economic feasibility, and control over the crop environment. Management factors that may be modified are crop rotation, plowing, irrigation schedule management, drainage, embankment work, soil improvement, grazing, pruning, thinning, harvesting, and controlling the amount and timing of fertilizer and pesticide use [12].

The APEX model was developed to apply the functions of the EPIC model to an entire farm or a small-scale watershed. This model is also equipped with the function of tracking water, sediments, nutrients, and pesticides that exist in various areas, such as complex terrains, watercourse systems, and watershed exits [10,12]. Using this tracking mechanism, the interactions among the surface runoff, return flow, sedimentation and decomposition of deposits, transfer of nutrients, and groundwater flow can be monitored. The evaluation enables the estimation of water quality that considers nitrogen (ammonia, nitrate, and organic form), phosphorus (soluble, adhesive, mineral, and organic), and pesticide concentrations [12].

BMPs are methods used to control pollutants that occur from NPSs to reduce their concentration levels and satisfy the standards of the target water quality. In the United States, numerous states, including New York and South Carolina, promulgate manuals that specify the actions and education required to apply BMPs. Furthermore, government agencies require forest and land owners and the forestry department to work together to implement BMPs [21].

P. Tuppad et al. [22] evaluated the effects of various BMPs, such as nutrient management, brush management, range planting, conservation cropping, contour farming, terrace, ponds, grade stabilization structures, and grassed waterways, using the APEX model on Mill Creek Basin (280 km^2), Texas.

2.2. Development of the QAPEX Analysis System

The QAPEX analysis system was developed using Python (www.python.org, accessed on 1 February 2018), based on the QGIS platform. Python exhibits considerable potential for future use because it is free, its grammar is easy, and it contains various libraries that include open-source packages, such as PyQT. The QAPEX analysis system uses a terrain analysis with digital elevation models (DEMs; TauDEM, http://hydrology.usu.edu/taudem/taudem5, accessed on 1 February 2018) [23], which offer numerous functions to process geographical information, and the QGIS software. The system is based on QSWAT developed by Dile et al. [19].

The QGIS software provides numerous functions to the QAPEX analysis system, including layer panels that indicate legend information and canvases that enable visualization for users. In particular, the QGIS software uses the geospatial data abstraction library (GDAL, www.gdal.org, accessed on 1 February 2018) to process data in the form of raster and shape. The GDAL supports 26 vector data formats, including the ESRI shapefile, and 72 raster file formats, including ArcInfo ASCII Grid and GeoTIFF. Because the QAPEX analysis system uses TauDEM, it requires a DEM in GeoTIFF format. If a raster file provided by a user is not in the GeoTIFF format, then GDAL may be used to convert file formats.

The main functions of the QAPEX analysis system are to (1) delineate watershed, (2) create hydrological landuse units (HLUs), (3) edit inputs and run APEX, and (4) visualize the output, as displayed in Figures 1 and 2. The first and second steps are performed using QSWAT functions developed by Dile et al. [19].

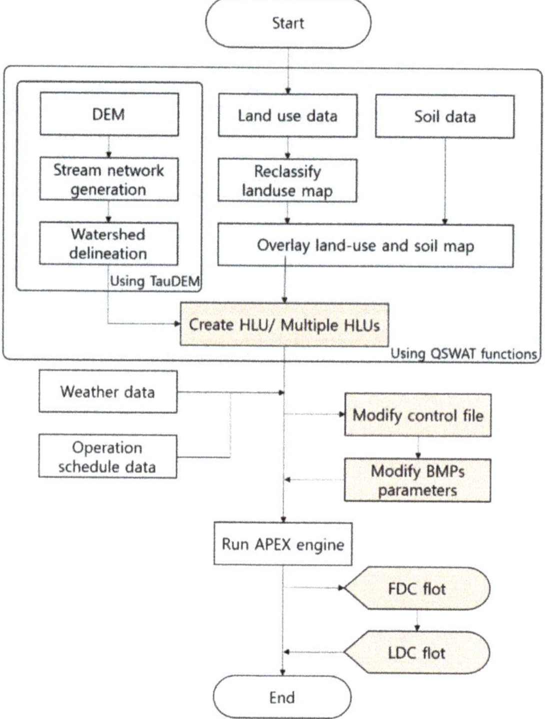

Figure 1. Flow chart of QAPEX analysis system [19].

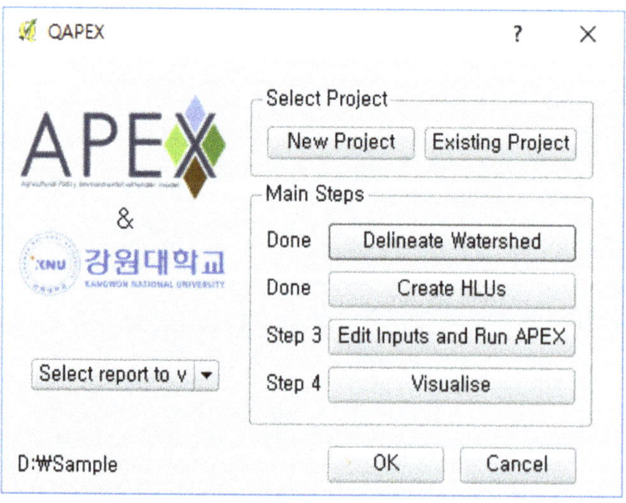

Figure 2. Major functions of the QAPEX analysis system.

2.2.1. Linking the Tool (Watershed Delineation and HLU Creation) from QSWAT to QAPEX Analysis System

The script used to estimate the watershed in QSWAT is also used in the QAPEX analysis system. The script uses various functions to estimate and delineate watersheds, such as the pit removal of TauDEM, calculation of the flow path and slope, estimation of the catchment area with the flow direction method, and delineation of a drainage network using the threshold value of a catchment area [19,23–25]. In TauDEM 5.1.2 version, the QAPEX analysis system is used for realizing the message passing interface (MPI), which allows the division of a single task into multiple processes [19,26]. However, to use the MPI function in the QAPEX analysis system, users must install the free MPI program offered by Microsoft.

The QAPEX analysis system is equipped with the QSWAT function of defining a stream using a stream file (vector file) held by users [19]. This function is useful for accurately identifying the location of a stream because it is difficult to do so with the TauDEM function, which exhibits a flat DEM. To form a stream network within the model, a threshold value (or the number of DEM cells) is required. In the QAPEX analysis system, the basic threshold value is set as 1% of the DEM size, and users can adjust the threshold value to either increase or decrease the number of streams [19]. To delineate a basin in TauDEM, at least one outflow gate defined with a point is required. In the QAPEX analysis system that uses QSWAT functions, users can use an outflow gate file (vector file) of a stream and specify outflow gates within the QGIS software canvas [19]. The APEX model creates models after homogeneous properties in territorial units of HLUs, such as weather, soil, land use (farming management schedule), and topographic maps [12]. However, as in the case of ArcAPEX, assigning HLUs to each subwatershed after delineating a watershed limits the simulation of the main HLU when multiple HLUs exist within a subwatershed. Therefore, this study applied hydrological response units (HRUs) used in the SWAT model to develop a method of simulating multiple HLUs within each subwatershed, instead of selecting a representative HLU from multiple HLUs. The "multiple HLUs within each subwatershed" method is useful because it can precisely reflect actual conditions in the model by considering HLUs that exclude the representative HLU (land use and soil information).

The QAPEX analysis system adopted the function used in the QSWAT to reduce the number of HRUs. This function can be categorized into two areas. In the first area, as in the case of the ArcAPEX model, we can assign the representative HLU within a watershed using either of the following steps: (1) selecting the largest value among land use, soil, and slope range or (2) selecting the largest HLU within a subwatershed. In the second area, the simulation of the "multiple HLUs within each subwatershed" method is performed by defining the number of subwatershed by (3) filtering the range of land use, soil, and slope; (4) filtering according to the surface area; or (5) directly designating a number [19]. This function enables users to select one of the five methods to automatically apply either an HLU or multiple HLUs.

2.2.2. Developement of Input Data Building Tool and BMP Application Tool in the QAPEX Analysis System

In this study, to run the QAPEX analysis system within the QGIS software, the database built into the APEX model was used to construct data files on crop properties, plowing properties, fertilizer properties, pesticide properties, soil properties, season information, and meteorological sites. In addition, a database using TauDEM was used to develop a data file (".SUB) on the properties of each HLU and watershed. In particular, the APEX-CONT.DAT file used to manage the operation of the QAPEX analysis system may be created based on the database built within the APEX model, and users can write variables for the new control file. In addition, users can input meteorological data and the farming management schedule required to run the QAPEX analysis system. The APEX model uses the engine of the WINAPEX0806 model, provided by Texas A&M AgriLife Research (https://epicapex.tamu.edu, accessed on 1 February 2018), at no charge to run the QAPEX analysis system.

Koo et al. [27] developed farming management schedules for beans, corn, potatoes, sweet potatoes, red beans, napa cabbage, white radishes, peppers, onions, green onions, garlic, spinach, lettuce, pumpkins, cabbage, cucumbers, watermelons, carrots, sesames, perillas, and peanuts. Furthermore, Choi et al. [11] devised farming management schedules for 25 regions in South Korea by considering each region's rice field characteristics.

As a model that was developed to analyze the effects of farmland management on small watersheds, the APEX model can consider not only various farming activities, such as drainage for irrigation, intertilling, buffer strips, fertilizer and manure use, and plowing, but also a variety of BMPs. The BMPs that can be applied in the APEX model are as follows: (1) structural preservation methods, such as check dams, diversion dikes, filter strips, grassed waterways, and interceptor swales/rain gardens; (2) nonstructural preservation methods, such as cropland conversion to pasture, no till, rainwater harvesting, and vegetation; and (3) waterway preservation methods, such as channel protection, riparian forest buffer, mulching, and stream restoration [28].

The interface to apply BMPs, however, is not available on the WinAPEX system and ArcAPEX interface, which causes inconvenience because of the manual search and modification of all variables related to BMPs. Furthermore, the variable-adjusting interface on the WinAPEX system simply lists numerous variables, which renders it difficult for users to determine certain variables from the list. Therefore, this study addressed this inconvenience by developing an interface within the QAPEX analysis system, such that variables related to BMPs can be easily adjusted.

2.2.3. Development of Visualization Function in QAPEX Analysis System

In 2004, South Korea first introduced total maximum daily loads (TMDLs), a scientific water quality management method that sets and manages the total discharge of contaminants by region, in the Nakdong River water system, and gradually expanded its application to water systems in the Geumgang, Yeongsangang, and Seomjingang rivers [29]. A commonly used method for planning TMDLs is schematizing an LDC to establish appropriate control methods according to flow duration conditions [30]. The LDC displays the relationship between the individual water quality and target water quality under the total flow condition of a stream [31]. The LDC can identify the effects of seasonal flow changes on the water quality and provides an easy understanding of the frequency and volume of target water quality and volume of allowed reduced loads. Therefore, investigating the cause of pollutants exceeding the target water quality is critical [32]. Numerous states in the United States use the LDC in TMDL setup, data analysis, and load management techniques for points and NPSs based on the flow size [33].

The results of the APEX model are generated through the variable set in PRNT****.DAT files. Furthermore, simulation results can be generated from a subwatershed or an entire watershed according to the day, month, and year; a summary file of a watershed can be obtained [10]. However, the simulation results from the existing APEX model are displayed in the text form, requiring users to perform additional study to visualize the results. The QAPEX analysis system includes the QGIS software, which enables users to visualize simulation results using the PyLab library of the Python programming language. This system incorporates an interface that can graph the information in *.RCH files for GIS-related specialized downstream areas. The interface is equipped with a general plot function that chronologically graphs simulation results by day, month, and year, and the FDC and LDC plot functions that graph simulation results on the FDC and LDC.

The LDC produced by the QAPEX analysis system follows the methods outlined by the NIER. (1) The FDC is created using flow data from the QAPEX analysis system results, (2) the target water quality input by users is converted into a target LDC, and (3) an LDC is produced for variables selected by users.

2.3. Demonstrative Application of the QAPEX Analysis System

This study compared the results obtained using the main HLU method and multiple HLU functions on the QAPEX analysis system. The QAPEX analysis system was applied to a model area to provide an example of the visualization function of the analysis system. The model area selected in this study is the Jaun-ri watershed located in Nae-myeon, Hongcheon-gun, Gangwon-do Province, South Korea. Along with the Mandae and Ga-a districts, Jaun district, which includes Jaun-ri, was re-designated as an NPS control area in October 2015. The NPS reduction project was performed in the Jaun-ri area, located in the upper Soyangho Lake (MOE, nonpoint.me.go.kr). The Jaun-ri watershed covered an area of 78.25 km^2 and its average elevation was 812.74 m, ranging between 570 and 1476 m.

The DEM used in the QAPEX analysis system to create a watershed and HLU/multiple HLUs is based on the system developed by Koo et al. [34], and it is processed into a 30 m $\times$ 30 m DEM. Koo et al. [34] previously produced a digital map on a 1:5000 scale into a 5 m $\times$ 5 m DEM suggested by Park et al. [35] by following DEM creation methods provided by the National Geographic Information Institute. The data to create an HLU/multiple HLUs used major classifications of land use maps provided by the MOE and detailed soil maps provided by the National Institute of Agricultural Science of Korea RDA. This study used data on soil characteristics according to each soil series produced by Koo et al. [27], which used information available on the Korean Soil Information System (soil.rda.go.kr). A farming management schedule for potatoes was applied to the farming area. Nonfarming areas, including built-up and dry areas, forest areas, grasslands, wetlands, bare land, and water bodies, used farming management schedules established in the ArcAPEX interface to simulate respective soil use. Furthermore, the variables used in Choi et al. [11] to simulate South Korea's environment in the APEX-Paddy development were used as input variables required to run the QAPEX analysis system.

3. Results

3.1. Comparing Characteristics According to the Classification Units of Subwatersheds

This study compared the characteristics between classification units (HLU/multiple HLUs) of the subwatershed by implementing the main HLU/multiple HLU methods of the QAPEX analysis system for the model area of Jaun-ri (Nae-myeon, Hongcheon-gun). Based on the analysis, 41 main HLUs were present in Jaun-ri, which were further divided into 821 multiple HLUs (Figure 3). Among the 41 main HLUs, 39 were forests (FRSD) and two were agricultural areas (AGRL), indicating that 99.92% of the area are forests and 0.08% are used for farming. The outcome revealed the phenomenon where forests are applied as the main HLUs within a subwatershed because the majority of land use in South Korea is defined as so. Through the implementation of the multiple HLU function, 0.7% of land use was for build-up and dry areas (URBN), 13.1% for agricultural areas (AGRL), 83.0% for forests (FRSD), 1.8% for grasslands (PAST), 1.0% for wetlands (WETN), 0.3% for bare land (AGRC), and 0.1% for water bodies (Table 1). The simulation outcomes are similar to those of actual land use areas. Therefore, in the case where multiple forms of land use exist within a subwatershed, creating uncertainties toward the implementation of the main HLU method, it is accurate to apply the multiple HLU method to replicate actual circumstances within the simulation model.

Table 1. Results of land use classification using HLU/multiple HLUs.

Function	URBN	AGRL	FRSD	PAST	WETN	AGRC	WATR
HLUs	-	0.06 (0.08)	78.19 (99.92)	-	-	-	-
Multiple HLU	0.58 (0.7)	10.26 (13.1)	64.96 (83.0)	1.43 (1.8)	0.75 (1.0)	0.20 (0.3)	0.08 (0.1)

unit: km^2 (%)

URBN: build-up and dry areas AGRL: agricultural areas FRSD: forests PAST: grasslands WETN: wetlands AGRC: bare land WATR: water.

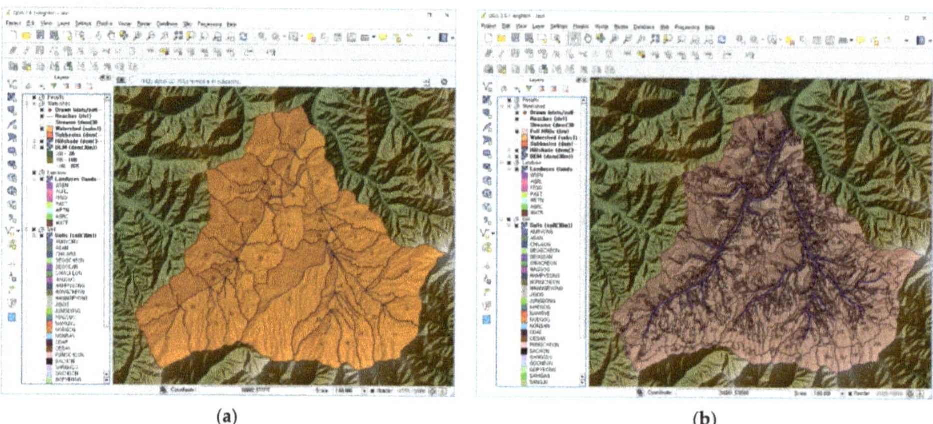

(a)   (b)

Figure 3. Results of HLU/multiple HLU function in the QAPEX analysis system for the study area (Jaun-ri). (**a**) HLU, (**b**) Full HLU (multiple HLUs).

3.2. Editing Database and Running the QAPEX Analysis System

3.2.1. Integrated Interface

An integrated interface (Figure 4) was developed to input and edit the data required to run the QAPEX analysis system in the QGIS software. The user first selected the database file of the APEX model, weather data from meteorological observation sites, and farming management schedule data through the integrated interface to run the QAPEX analysis system. The user then created a control file (APEXCONT.DAT) to manage the operation of the system. The control file can be created using two methods: applying variables entered into the APEX model database and entering and applying new variables through the control file variable adjustment interface. After following the aforementioned procedure, data files necessary for running the QAPEX analysis system were created by clicking on the "Apply" button displayed in Figure 4.

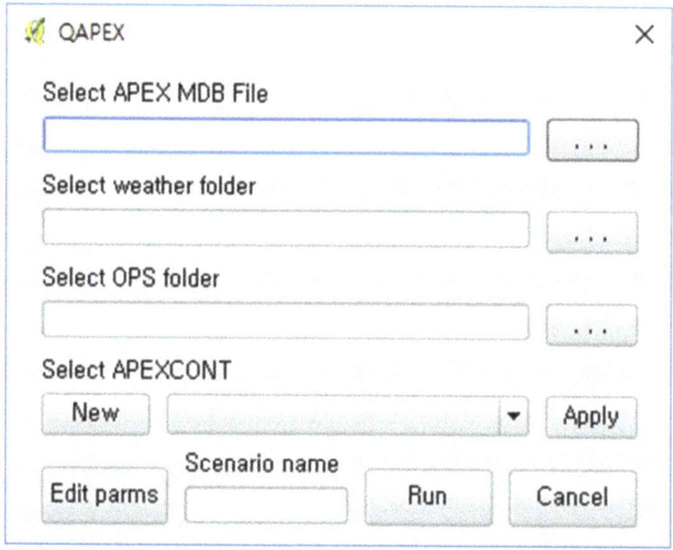

Figure 4. User interface for connecting the database of the APEX model and running QAPEX.

The "Edit parms" function on the integrated interface leads users to an interface where they can edit data files created through the integrated interface, such as the watershed variable adjustment interface and BMP interface of the QAPEX analysis system. Finally, after naming the file and saving the folder through the "Scenario name," the "Run" button is clicked to run and save the QAPEX analysis system.

3.2.2. Control File Variable Adjustment Interface

The control file (APEXCONT.DAT) used to adjust the variables necessary to run the APEX model was developed so that it could be created in two methods. The first method involves fetching variables entered in the APEX model database, and the second method is the user creating the APEXCONT.DAT file by adjusting variables using the control file variable adjustment interface (Figure 5). In the second method, the initial set points of the entered variables were set as the default values of the WinAPEX system. The interface to adjust variables in the control file was organized in the same layout (order and line) as the APEXCONT.DAT file for user convenience.

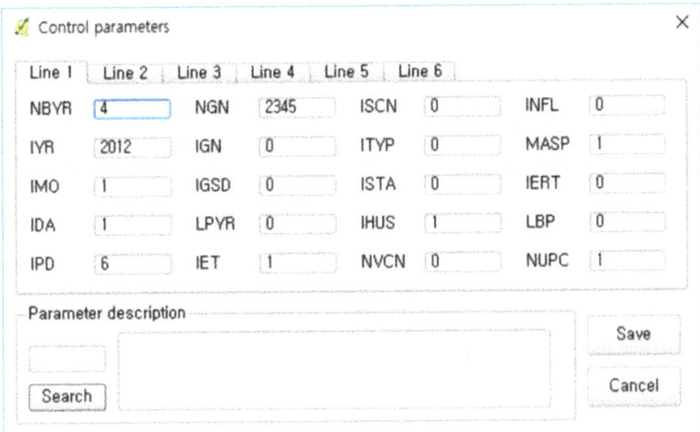

Figure 5. User interface for creating and editing the control file (APEXCONT.DAT).

Furthermore, to have access to the explanations of variables within the QAPEX analysis system, "Parameter description" was added to the interface so that users could read descriptions on a searched variable.

3.2.3. Watershed Variable Adjustment Interface

Similar to the case of QSWAT using TauDEM to derive the characteristics of a watershed, the QAPEX analysis system applies computed results of TauDEM, HLU/multiple HLUs, and weather and farming management schedule data entered by users to develop a data file (*.SUB) on the characteristics of a watershed, including soil characteristics (INPS), farming management schedule (IOPS), meteorological sites (IWTH), latitude (YCT) and longitude (XCT), watershed surface area, waterway length from the farthest point away from the watershed exit (CHL), waterway depth (CHD), inclination of main waterways (CHS), waterway length of tracking downstream area (RCHL), and waterway slope of tracking downstream area (RCHN). The initial set points for the moisture content in the deposited snow (SNO), residues in dead crops (STDO), number of manning related to fields (UPN), and irrigation code (NIRR) applied in *.SUB used default values in the WinAPEX model.

The SUB file created through the QAPEX analysis system may be modified to fit the characteristics of relevant watersheds by users through the watershed variable adjustment interface. The user selects the subwatershed number to edit and its appropriate HLU and clicks on the "Read" button (Figure 6), and the interface reveals the HLU with applied variables. The user may adjust the necessary variables and click on "Save" to edit the variables of the HLU. Furthermore, to change the same variable for the entire watershed area, select "All" from the "Select subbasin" combo box to edit the variable for all HLUs. The changes are applied to all variables, except those marked "-".

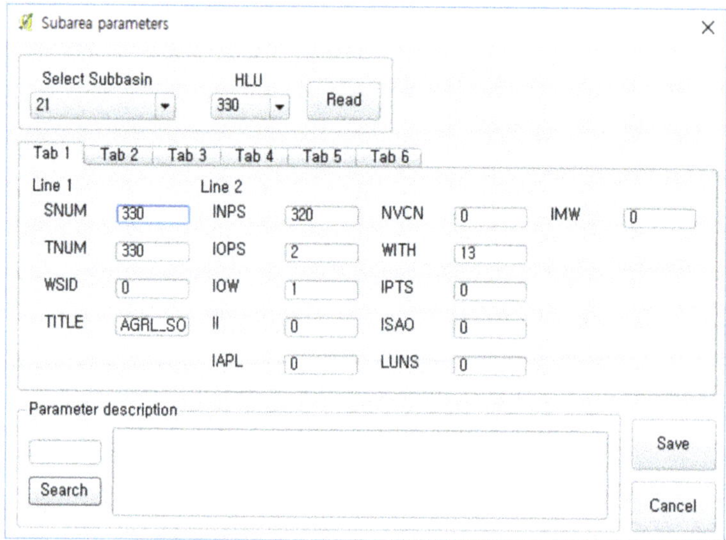

Figure 6. User interface for adjusting the subbasin (*.SUB)—related parameters.

The description of variables displayed in the interface can be verified by typing the name in "Parameter description." Two lines are organized per tab in the interface, such that the variables used in the *.SUB file can be easily located.

3.2.4. BMP Variable Adjustment Interface

This study developed a user interface (Figure 7) that can adjust variables related to BMPs within the QAPEX analysis system for user convenience. Among the BMPs that are considered in the APEX model, this study first applied BMPs on structural installation: check dams, diversion dikes, filter strips, grade stabilization structures, grassed waterways, interceptor swales/rain gardens, pipe slope drains, sediment basins, silt fences, terraces, triangular sediment dikes, and wetland creation. When users select the HLU and BMP of a subwatershed through the interface, the variables relevant only to the watershed appear on the right, and they may adjust the variables to consider the BMP.

3.3. Development of the Visualization Function Interface

3.3.1. Integrated Interface

This study developed an integrated interface (Figure 8) to provide users with functions to graph the general plot, FDC plot, and LDC plot. To use the visualization function in the QAPEX analysis system, users must set the scenario, subbasin, HLU, and graph period to be displayed on the graph, and then enter relevant data for each function. Next, clicking on "Plot" in Figure 8 produces a graph for each function. The "Form" function is added to allow users to change the layout of graphs, such as the inclusion of the auxiliary axis, inclusion of a legend and its location, line style and color, axis titles, unit of the *Y*-axis, and graph title.

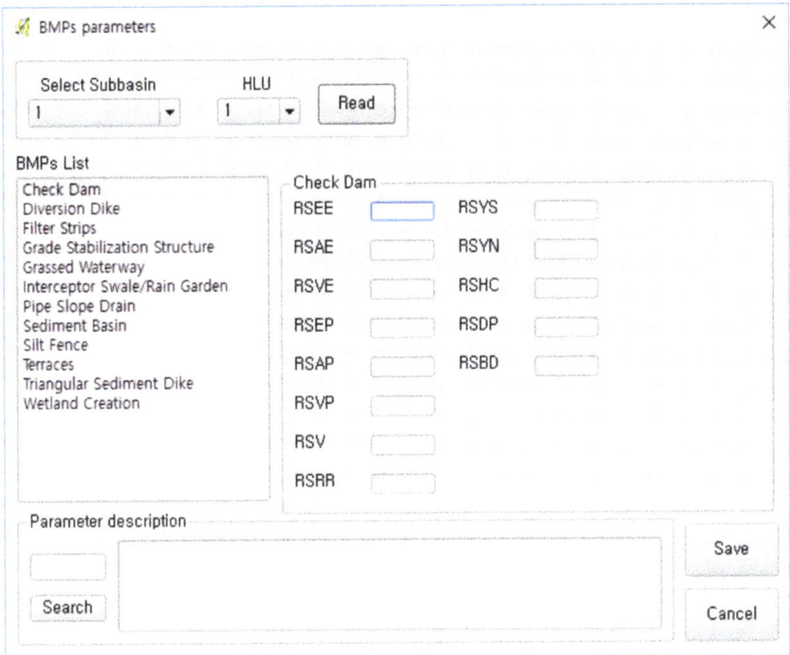

Figure 7. User interface for adjusting BMP-related parameters.

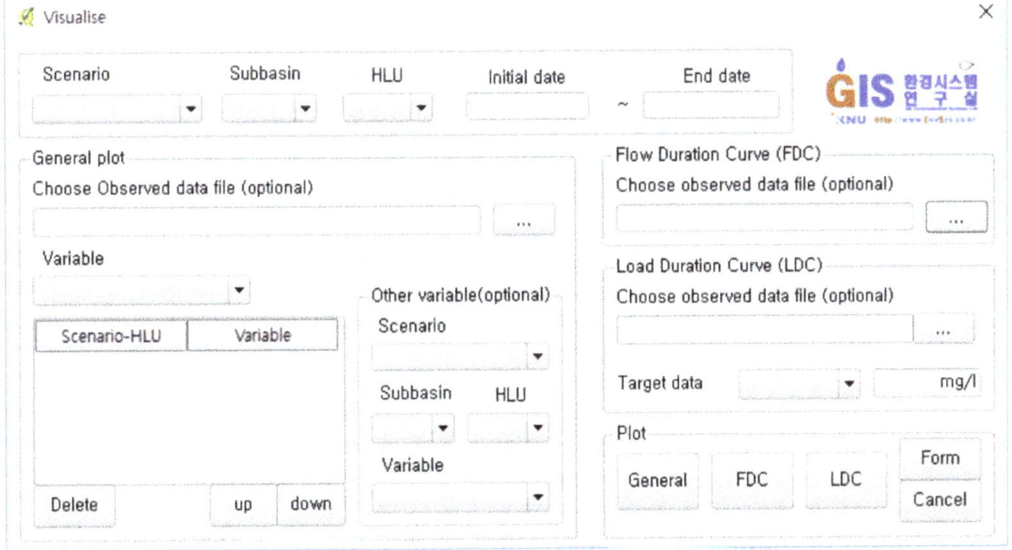

Figure 8. User interface for the visualization of results in the QAPEX analysis system.

3.3.2. General Plot

The "general plot" function in the QAPEX analysis system graphs simulation results in a chronological manner of day, month, and year. This function can selectively visualize the variables generated in the *.RCH file yielded by the simulation results of the QAPEX analysis system. Not only can this system graph a number of variables simultaneously, but it can also retrieve *.RCH files from other scenarios and visualize two or more scenarios simultaneously. Furthermore, by entering csv-format observation data, the "general plot" function can plot the observation data and simulation results within the same graph. Therefore, not only is it possible to visualize simulation results through the "general plot" function, it is also possible to perform various analyses, such as comparing scenarios, observation data, and simulation results.

Figure 9 displays the application of the "general plot" function on the model area (Jaun-ri) for the water yield outflow (CMS). The layout of the graph is set at default values, which may be changed by users through the "Form" function to edit the graph title, axis titles, legend, variable names, and units.

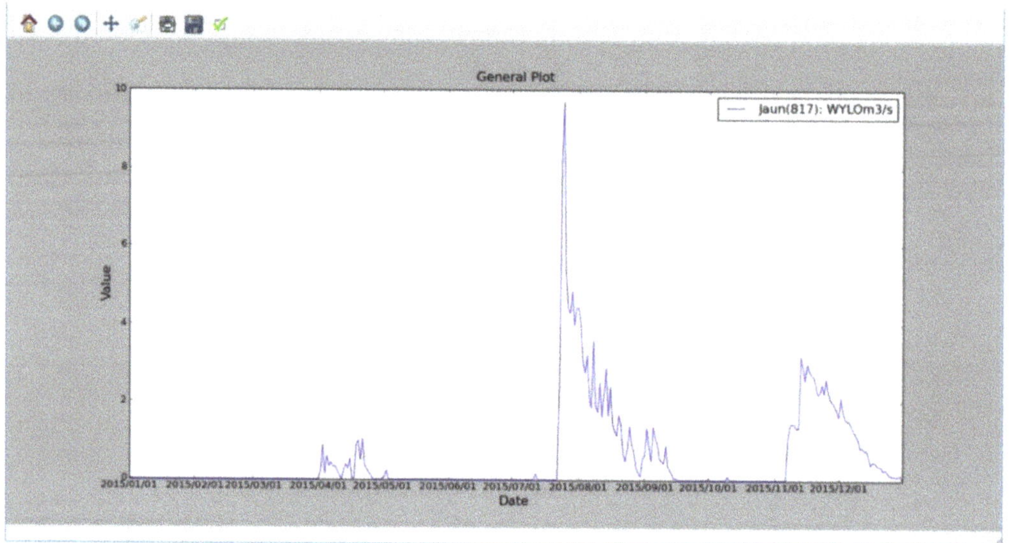

Figure 9. General plot function among the visualization results of the QAPEX analysis system for the water yield outflow (CMS).

3.3.3. FDC/LDC Plot

The QAPEX analysis system is equipped with the functions "FDC plot" and "LDC plot" to display the simulation results (*.RCH) for the FDC and LDC, respectively. The "FDC plot" function can be run without an additional data input. The LDC plot function requires users to select the variables and target water quality to be demonstrated through the LDC. By adding the input data of the observation date, CMS, and observation data in csv format, the FDC/LDC plot functions produce FDC/LDC only composed of the observation data in addition to FDC/LDC generated by simulation results. Because the FDC/LDC plot functions developed in this study can indicate FDCs and LDCs of areas that partially or completely lack observation data, the two functions can compensate for the absence of monitoring data from temporal and locational limitations.

Figure 10 displays the application of "FDC plot" and "LDC plot" functions for the model area (Jaun-ri), where the LDC plot reveals the results for nitrogen (NO_3O, kg) transferred with water out of the downstream area. The default values were used for the layout of the graph.

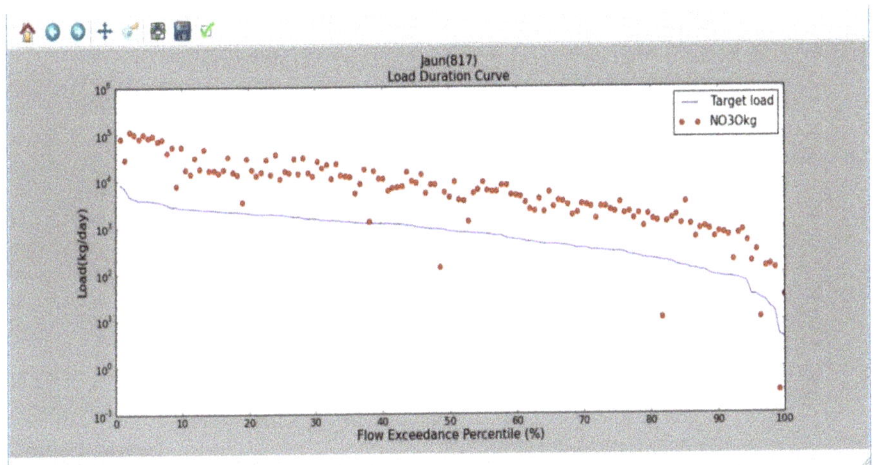

Figure 10. LDC plot function among the visualization results of the QAPEX analysis system for the study area (Jaun-ri).

4. Conclusions

Based on the EPIC model, the APEX model was improved and expanded into various models, such as WinAPEX, ArcAPEX, and i_APEX. This study synced the APEX model with the QGIS software, an open-source GIS software, and developed the QAPEX analysis system that can use GIS functions without a fee. The study also reduced the inconvenience of users manually inputting data of watersheds, as in the case of the WinAPEX system. Furthermore, the system created the "multiple HLUs within each subwatershed" method to consider multiple land uses and soil maps, which is an improvement on the ArcAPEX interface where a subwatershed is calculated as a single land use/soil map through the main HLU method. The "multiple HLUs within each subwatershed" method eliminates uncertainties resulting from single land use/soil map simulation. This study also developed an interface to adjust variables for BMPs that are performed as control methods for NPSs so that BMPs may be applied and run in the model. Furthermore, the visualizations of the comparison between the simulation results of multiple scenarios and the functions of FDC/LDC were added. Such visualization functions express the water quality and water yield outflow through graphs, rendering it easy for users to understand. In particular, the LDC is advantageous for use in identifying the cause of pollutants exceeding the target water quality. Therefore, the QAPEX analysis system can be used as a tool to gather national policy data by analyzing NPS control areas or reducing the effects of applying BMPs.

Author Contributions: Conceptualization, J.K. (Jayoung Koo) and J.K. (Jonggun Kim); Data curation, J.J.; Formal analysis, J.K. (Jayoung Koo) and M.-K.K.; Funding acquisition, K.-J.L.; Investigation, J.K. (Jayoung Koo) and S.L.; Methodology, J.K. (Jayoung Koo), D.-S.S. and S.L.; Project administration, K.-J.L.; Resources, J.K. (Jayoung Koo) and J.J.; Supervision, K.-J.L.; Validation, J.K. (Jayoung Koo), J.K. (Jonggun Kim) and J.R.; Visualization, J.K. (Jayoung Koo); Writing—original draft, J.K. (Jayoung Koo); Writing—review & editing, J.K. (Jayoung Koo), J.K. (Jonggun Kim) and S.L. All authors have read and agreed to the published version of the manuscript.

Funding: This research was supported by the Korea Environment Industry and Technology Institute (KEITI) through the Aquatic Ecosystem Conservation Research Program funded by the Korean Ministry of Environment (MOE), grant number 2020003030004.

Institutional Review Board Statement: Not applicable.

Informed Consent Statement: Not applicable.

Data Availability Statement: QAPEX analysis system described in the manuscript will be downloaded from https://github.com/Bong67/QAPEX (accessed on 1 February 2018).

Acknowledgments: The authors gratefully acknowledge contributions from the Ministry of Environment of Korea for supporting this work. Finally, we would like to thank the anonymous reviewers, whose comments and suggestions helped improve the manuscript.

Conflicts of Interest: The authors declare no known competing financial interests or personal relationships that could have influenced the work reported in this paper.

References

1. Lee, S.H.; Heo, I.H.; Lee, K.M.; Kim, S.Y.; Lee, Y.S.; Kwon, W.T. Impacts of climate change on phenology and growth of crops: In the case of Naju. *J. Korean Geogr. Soc.* **2008**, *43*, 20–35. (In Korean)
2. Kim, M.; Kwon, S.; Jung, G.; Hong, S.; Chae, M.; Yun, S.; So, K. Small-Scale Pond Effects on Reducing Pollutants Load from a Paddy Field. *Korean J. Environ. Agric.* **2013**, *32*, 355–358. [CrossRef]
3. Choi, J.Y. *Study on the Establishment of Designation Criteria for Nonpoint Pollution Source Management Area*; Ministry of the Environment: Sejong, Korea, 2007. (In Korean)
4. Ministry of the Environment (MOE). *Management Measures of Nonpoint Source Pollutants in the Watershed of Lake Soyang*; Ministry of the Environment: Sejong, Korea, 2007; (In Korean). Available online: http://www.me.go.kr/home/web/policy_data/read.do?pagerOffset=0&maxPageItems=10&maxIndexPages=10&searchKey=title&searchValue=%EC%86%8C%EC%96%91%ED%98%B8&menuId=92&orgCd=&seq=878 (accessed on 1 June 2018).
5. Related Ministries Joint. *The 3rd (2021–2025) Rainfall Runoff Nonpoint Pollution Source Management Comprehensive Measures*; Ministry of the Environment: Sejong, Korea, 2020; (In Korean). Available online: http://eng.me.go.kr/home/web/policy_data/read.do?pagerOffset=70&maxPageItems=10&maxIndexPages=10&searchKey=&searchValue=&menuId=10259&orgCd=&condition.toInpYmd=null&condition.fromInpYmd=null&condition.deleteYn=N&condition.deptNm=null&seq=7677 (accessed on 1 December 2021).
6. Choi, J.Y.; Park, B.K.; Kim, J.S.; Lee, S.Y.; Ryu, J.C.; Kim, K.H.; Kim, Y.S. Determination of NPS pollutant unit loads from different landuses. *Sustainability* **2021**, *13*, 7193. [CrossRef]
7. National Institute of Environmental Research (NIER). *A Monitoring and Management Scheme for the Nonpoint Sources (III)*; National Institute of Environmental Research: In-Cheon, Korea, 2015. (In Korean)
8. Srinivasan, R.; Arnold, J.G. Integration of a basin-scale water quality model with GIS. *J. Am. Water Resour. Assoc.* **1994**, *30*, 453–462. [CrossRef]
9. Bicknell, B.R.; Imhoff, J.C.; Kittle, J.L.; Donigian, A.S.; Johanson, R.C. *Hydrological Simulation Program-FORTRAN, User's Manual for Release 11*; Environmental Protection Agency 600/R-97/080; United States Environmental Protection Agency: Athens, GA, USA, 1997.
10. Kim, M.K.; Choi, S.K.; Jung, G.B.; Kim, M.H.; Hong, S.C.; So, K.H.; Jeong, J.H. APEX (agricultural policy/environmental eXtender) model: An emerging tool for agricultural environmental analyses. *Korean J. Soil Sci. Fertil.* **2014**, *47*, 187–190. [CrossRef]
11. Choi, S.C.; Kim, M.K.; So, K.H.; Jang, T.I. Application of APEX-PADDY model considering rice cultivation environment. *Rural Resour.* **2017**, *58*, 23–27. (In Korean)
12. Williams, J.R.; Izaurralde, R.C.; Steglich, E.M. *Agricultural Policy/Environmental eXtender Model Theoretical Documentation*; Agrilife Research Texas A&M System: Temple, TX, USA, 2012.
13. National Institute of Agricultural Sciences (NAS). *Agricultural Policy/Environmental eXdenter Model User's Manual*; Version 0806; National Institute of Agricultural Sciences: Wanju, Korea, 2015. (In Korean)
14. Williams, J.R.; Izaurralde, R.C. Chapter 18. The APEX model. In *Watershed Models*; Singh, V.P., Frevert, D.K., Eds.; CRC Press: Boca Raton, FL, USA; Taylor & Francis Group: Oxfordshire, UK, 2005; pp. 437–482.
15. Tuppad, P.; Winchell, M.F.; Wang, X.; Srinivasan, R.; Williams, J.R. ArcAPEX: ArcGIS interface for agricultural policy environmental extender (APEX) hydrology/water quality model. *Int. Agric. Eng. J.* **2009**, *18*, 59.
16. Feng, Q.; Engel, B.A.; Flanagan, D.C.; Huang, C.; Yen, H.; Yang, L. Design and development of a web-based interface for the agricultural policy environmental eXtender (APEX) model. *Environ. Model. Softw.* **2019**, *111*, 368–374. [CrossRef]
17. National Disaster Management Institute (NDMI). *A Study on Application Method of Smart CCTV Based on 3D-Map for Disaster Management*; National Disaster Management Institute: Ulsan, Korea, 2014. (In Korean)
18. Chen, D.; Shams, S.; Carmona-Moreno, C.; Leone, A. Assessment of open source GIS software for water resources management in developing countries. *J. Hydro-Environ. Res.* **2010**, *4*, 253–264. [CrossRef]
19. Dile, Y.T.; Daggupati, P.; George, C.; Srinivasan, R.; Arnold, J. Introducing a new open source GIS user interface for the SWAT model. *Environ. Model. Softw.* **2016**, *85*, 129–138. [CrossRef]
20. Williams, J.R.; Jones, C.A.; Dyke, P.T. A Modeling Approach to Determining the Relationship between Erosion and Soil Productivity. *Trans. ASAE* **1984**, *27*, 0129–0144. [CrossRef]
21. Choi, H.T.; Kim, S.J.; Bae, S.O. *Best Management Practices for Water Quality Conservation in Forest Basin*; National Institute of Forest Science: Seoul, Korea, 2012; (In Korean). Available online: http://know.nifos.go.kr/book/search/DetailView.ax?&cid=160422 (accessed on 1 February 2018).

22. Tuppad, P.; Santhi, C.; Wang, X.; Williams, J.R.; Srinivasan, R.; Gowda, P.H. Simulation of conservation practices using the APEX model. *Appl. Eng. Agric.* **2010**, *26*, 779–794. [CrossRef]
23. Tarboton, D.G. A new method for the determination of flow directions and upslope areas in grid digital elevation models. *Water Resour. Res.* **1997**, *33*, 309–319. [CrossRef]
24. Tesfa, T.K.; Tarboton, D.G.; Watson, D.W.; Schreuders, K.A.T.; Baker, M.E.; Wallace, R.M. Extraction of hydrological proximity measures from DEMs using parallel processing. *Environ. Model. Softw.* **2011**, *26*, 1696–1709. [CrossRef]
25. Wilson, J.P. Digital terrain modeling. *Geomorphology* **2012**, *137*, 107–121. [CrossRef]
26. Wu, Y.; Li, T.; Sun, L.; Chen, J. Parallelization of a hydrological model using the message passing interface. *Environ. Model. Softw.* **2013**, *43*, 124–132. [CrossRef]
27. Koo, J.Y.; Kim, J.G.; Choi, S.K.; Kim, M.K.; Jeong, J.H.; Lim, K.J. Construction of database for application of APEX model in Korea and evaluation of applicability to highland field. *J. Korean Soc. Agric. Eng.* **2017**, *59*, 89–100. (In Korean) [CrossRef]
28. Waidler, D.; White, M.; Steglich, E.; Wang, S.; Williams, J.; Jones, C.A.; Srinivasan, R. *Conservation Practice Modeling Guide for SWAT and APEX*; Texas Water Resources Institute: College Station, TX, USA, 2011; Available online: https://hdl.handle.net/1969.1/94928 (accessed on 1 February 2018).
29. Park, J.D.; Kim, J.L.; Rhew, D.H.; Jung, D.I. A study on the water quality patterns of unit watersheds for the management of TMDLs in Nakdong River Basin. *J. Korean Soc. Water Qual.* **2010**, *26*, 279–288.
30. Yun, S.Y.; Ryu, J.N.; Oh, J.I. Water quality management measures for TMDL unit watershed using load duration curve. *J. Korean Soc. Water Wastewater* **2013**, *27*, 429–438. [CrossRef]
31. United States Environmental Protection Agency (USEPA). *An Approach for Using Load Duration Curves in the Development of TMDLs*; 841-B-07-006; Environmental Protection Agency: Washington, DC, USA, 2007.
32. Hwang, H.S.; Woon, C.G.; Kim, J.T. Application load duration curve for evaluation of impaired watershed at TMDL unit watershed in Korea. *J. Korean Soc. Water Qual.* **2011**, *26*, 903–909.
33. Babbar-Sebens, M.; Karthikeyan, R. Consideration of sample size for estimating contaminant load reductions using load duration curves. *J. Hydrol.* **2009**, *372*, 118–123. [CrossRef]
34. Koo, J.Y.; Yoon, D.S.; Lee, D.J.; Han, J.H.; Jung, Y.H.; Yang, J.E.; Lim, K.J. Effect of DEM resolution in USLE LS factor. *J. Korean Soc. Water Environ.* **2016**, *32*, 89–97. (In Korean) [CrossRef]
35. Park, C.S.; Lee, S.K.; Suh, Y.C. Development of an automatic generation methodology for digital elevation models using a two-dimensional digital map. *J. Korean Assoc. Geogr. Inf. Stud.* **2007**, *10*, 113–122. (In Korean)

MDPI

St. Alban-Anlage 66

4052 Basel

Switzerland

Tel. +41 61 683 77 34

Fax +41 61 302 89 18

www.mdpi.com

Sustainability Editorial Office

E-mail: sustainability@mdpi.com

www.mdpi.com/journal/sustainability